S. Chandrasekhar

The Man Behind the Legend

S. Chandrasekhar
The Man Behind the Legend

Editor

Kameshwar C. Wali

Syracuse University

ICP

Imperial College Press

Published by

Imperial College Press
516 Sherfield Building
Imperial College
London SW7 2AZ

Distributed by

World Scientific Publishing Co. Pte. Ltd.
P O Box 128, Farrer Road, Singapore 912805
USA office: Suite 1B, 1060 Main Street, River Edge, NJ 07661
UK office: 57 Shelton Street, Covent Garden, London WC2H 9HE

British Library Cataloguing-in-Publication Data
A catalogue record for this book is available from the British Library.

S. CHANDRASEKHAR: THE MAN BEHIND THE LEGEND

ISBN 1-86094-038-2

Printed in Singapore.

To Lalitha Chandrasekhar

CONTENTS

Introduction

Kameshwar C. Wali

> The simple is the seal of the true. And beauty is the splendor of truth.
>
> — *Chandra, at the conclusion of his Nobel Lecture, December 8, 1983*

Subrahmanyan Chandrasekhar, known simply as Chandra in the scientific world, is one of the foremost scientists of this century. With his prolific and wide-ranging contributions to physics, astrophysics and applied mathematics, he became and continues to be a legendary figure. However, in spite of his extraordinary accomplishments and tremendous success in contributing to our understanding of the evolutionary stages of stars and their ultimate destinies, to the vast landscape of Einstein's general relativity, and to the study of black holes, Chandra was never a household name like Feynman or Einstein. He remained a highly private individual, unknown to the world at large. The man behind the legend remained hidden. His almost ascetic, highly disciplined, organized and simplified life built around a single-minded pursuit of science made him seem unapproachable. Only a small number of people among his associates, students, admirers and close relatives were able to penetrate the seemingly impenetrable barrier. In commemorating him, I have drawn upon such people, asking them to write, not so much about Chandra's scientific triumphs, but more about Chandra as a person. In the articles that follow, you will see Chandra's rich and multi-faceted personality come alive through their encounters, anecdotes and reminiscences. For my part, I would like to offer a brief overview of Chandra's life and some of my reflections.

Chandra was born on October 19, 1910 (19–10–1910, as Chandra was fond of saying), in Lahore, Pakistan (then a part of colonial British India). His father, C.S. Ayyar, was in the government service, serving as the Deputy

Auditor-General of the Northwestern Railways. Chandra was the first son and the third child in a family of four sons and six daughters. His mother was a woman of great talent and intellectual attainments. Intensely ambitious for her children, she played a pivotal role in Chandra's pursuit of a career in pure science. Chandra's early education was at home; he was taught by his parents and private tutors. When he was twelve, his family moved to Madras. It was then that he began his regular schooling in the Hindu High School, Triplicane, during the years 1922–25, followed by his university education at the Presidency College, Madras. He graduated with a B.Sc. (Hons.) degree in 1930. Lalitha Doraiswamy, his future wife, was a fellow undergraduate studying physics. An exceptionally brilliant student throughout his student career, Chandra became nationally known by presenting an original research paper in physics as an undergraduate during the Indian Science Congress Meetings in 1929. Before he even graduated, he learned that he was awarded a Government of India Scholarship to continue his studies in England. By then he had established contact with R.H. Fowler in Cambridge which led him to decide to work under the latter's guidance for his doctoral degree. While all this was good news for a bright future, he faced at home a mother with a serious illness. The thought that, if he went to England, he might never see her again haunted him, and almost made him decide against going. However, his mother's insistence and persuasion, and her desire not to stand in the way of his future ultimately prevailed and he left for England in July of 1930. Unfortunately, his mother did not survive long enough to witness Chandra's successful career. She died in May of 1931, less than a year after Chandra's arrival in Cambridge.

The years from 1910 to 1930, the early years of Chandra in India, were turbulent ones. It was a time when India was struggling for independence from the British. At the same time British education was becoming nationally and internationally very influential in several fields, especially the sciences. While Gandhi, Nehru, Sarojini Naidu, Sardar Vallabhabhai Patel and other political leaders were becoming household names, there were men like Srinivasa Ramanujan, Meghanad Saha, Jagadish Chandra Bose, Satyendra Nath Bose, C.V. Raman and Rabindranath Tagore, who by their accomplishments in the sciences and arts had captured the imagination of the young, college-educated Indians. Their recognition by the British and the West was a source of inspiration and pride for millions of Indians who struggled in poverty but had dreams, although there were only a few who could transform them into reality. Chandra was among the fortunate few who did not have to struggle in poverty.

Gifted as he was, he developed a passion for mathematics and physics early in his life. He was hardly ten years old when he heard about the great Indian mathematician Srinivasa Ramanujan, who was to be a source of great inspiration and a role model for the rest of his life. Being the young nephew of the famous physicist Sir C.V. Raman opened the way for Chandra to get to know the senior physicists of India at the time and to experience the excitement of research. He came to know K.S. Krishnan, the principal collaborator of Raman in the discovery of the Raman effect, and through him learned about the visit of Arnold Sommerfeld to India and Madras in the fall of 1928. This led Chandra to a meeting with Sommerfeld and an introduction to the new quantum mechanics and Fermi–Dirac statistics. Chandra described his meeting with Sommerfeld as "the single most important" event in his early life which launched him on a research career (for details, see *Chandra: A Biography of S. Chandrasekhar*, University of Chicago Press, 1991, pp. 61–62).

Chandra's six years at Cambridge, England, from 1930 to 1936, first as a graduate student for three years and then as a Fellow of Trinity College, were years in which great discoveries in physics took place. They were also years of continuing British dominance in science. In 1930, on his long voyage from India to England, as a result of musing and calculations, Chandra arrived at a startling conclusion that subsequently became known as the celebrated discovery of the Chandrasekhar limit — the limit on the mass of a star that could become a white dwarf in its terminal stage. After years of hard work at developing, refining and putting the idea on a rigorous basis, Chandra presented his results at the January 1935 meeting of the Royal Astronomical Society. His findings raised challenging, fundamental questions. What happens to the more massive stars as they continue to collapse? Are there terminal stages of stars other than that of white dwarfs? Instead of getting appreciation and recognition for a fundamental discovery, Chandra unexpectedly faced what amounted to a public humiliation. Because, no sooner had he presented his paper than Sir Arthur Eddington, who had been a mentor and had followed the work closely, ridiculed the basic idea on which Chandra's conclusion was based. Eddington made it look like Chandra had gotten it all wrong. In the dramatic and unexpected encounter, young Chandra was pitted against an older, established, internationally renowned scientist whose authority, prestige and fame carried the day. More than two decades passed before the Chandrasekhar limit became an established fact and assumed its importance in astrophysical research, leading to further developments on other terminal stages of stars such as neutron stars and black holes.

In the face of this devastating experience, it was to the credit of young Chandra that he made a wise decision. Instead of a dogged fight and confrontation, he gracefully withdrew from the controversy. He stopped publishing further work in the theory of white dwarfs, and went on to research in a different area after summarizing his account in his first monograph (*An Introduction to Stellar Structure*, University of Chicago Press, Chicago, 1939; reprinted, Dover Publications, New York, 1967). This marked the beginning of Chandra's distinctive style of research throughout the rest of his life, of seeking his own perspective, and striving for a complete understanding of a particular area. When he felt he had accomplished it to his satisfaction, he presented it in a treatise or monograph; and then he abandoned the area entirely, never to return, without even maintaining a passing interest. He became inward-bound, seeking beauty and harmony in the workings of nature as exhibited in his mathematical equations.

In 1937, Chandra and Lalitha, newly married, arrived in the United States, to take up a research position at Yerkes Observatory of the University of Chicago, situated in Williams Bay, Wisconsin. Chandra immediately took on the task of developing a graduate program in astronomy and astrophysics. It wasn't too long before Chandra's reputation as a teacher, his youth and his enthusiasm for research began to attract students from all parts of the world: Paul Ledoux from Belgium; Mario Schönberg, Jorge Sahade and Carlos Cesco from Argentina; Gordon W. Wares, Ralph E. Williamson, Wasley S. Krogdahl, Margaret Kiess Krogdahl and Louis R. Henrich from the United States. These were some of Chandra's students and associates during his early years at Yerkes. Later on, after the Second World War, Guido Munch, Arthur D. Code, Donald E. Osterbrock, Esther Conwell, Jeremiah P. Ostriker and many others were to follow. In all, over fifty of them got their doctorate degree under his guidance and several of them have distinguished themselves as eminent scientists and statesmen of science. As a teacher and a lecturer, Chandra was a grand master who brought elegance and scholarship, which literally charmed the audience and kept it spellbound. The story is often told of how he drove some hundred miles between Yerkes Observatory in Williams Bay, Wisconsin, and the University of Chicago, week after week, to teach a class of two students. The administrators of the University of Chicago were not too happy with this allocation of the faculty resources; but when the whole class of Messrs. Lee and Yang got the Nobel Prize a few years later, they were quick to claim that to be an educational tradition unique to the University of Chicago. Chandra was also the sole editor of *The Astrophysical Journal* during the years 1952–1971. When he began, the journal was essentially the private property of the University of Chicago. Chandra played a decisive role in transforming it into the national journal of the American Astronomical Society and the foremost journal of its kind in the world.

The onset of the Second World War brought upon Chandra great anguish and concern for India. When Holland, Belgium, Norway and France went down to Germany in quick succession, and Greece and Turkey were trampled on, he was alarmed since he felt the security of India was threatened. However, Indian leaders felt it was the right moment to press for India's independence. Mahatma Gandhi initiated the "Quit India" movement, which led immediately to the outlawing of the Congress Party and the imprisonment of all its popular leaders, including Gandhi and Nehru. Strikes, riots and violence spread all over India. In spite of the news of these disturbing, unhappy and painful developments in India, and in spite of being a staunch supporter of India's independence, Chandra felt strong sympathy for Britain and a strong conviction that the world, including India, faced the worst danger if the Axis powers were victorious. Lalitha and Chandra participated in the civilian effort to help the British. When Japan struck at Pearl Harbor and America entered the war, Chandra joined the war effort and became a part of an outstanding group of scientists working at the Aberdeen Proving Grounds in Maryland. The group included John von Neumann, Ronald Gurney, Joseph Myer, L.H. Thomas, Martin Schwarzschild, Edwin Hubble, Robert Sachs and many others.

After the war, Chandra became more closely associated with the physics department of the University of Chicago, as he developed close collaboration with Enrico Fermi, and his researches took on new turns oriented more towards physics than astrophysics. Lalitha and Chandra subsequently moved from Williams Bay and made Chicago their permanent home. Chandra continued active research till the end of his life. After the publication of his book *The Mathematical Theory of Black Holes* in 1983, while continuing his researches on colliding gravitational waves, radial and non-radial oscillations of stars, he got interested in Newton's *Principia*. From a set of lectures, it got into the shape of a mathematical tome of some 400 pages with numerous illustrations. This was to be his last book. It came out just two months before his death. During the last few years of his life, it often appeared as though he was living for the completion of this monumental work. I also should mention that in the 1980s, his other abiding interest was getting a bust of Srinivasa Ramanujan made and presenting it to Janaki Ammal, the widow of Ramanujan. The full story of this episode is well told in Professor Askey's article.

The years I spent working with Chandra and writing his biography were the most enjoyable and creative years in my life. At the beginning, Chandra, as to many others, represented an awesome figure. I had this image of him as a glorious scientist-in-exile from India, his native country (mine as well). When I was a graduate student at the University of Wisconsin at Madison, Wisconsin, I came to know several people who knew him well. Among them was Carl Sagan, who was then a graduate student at Yerkes Observatory. My thesis advisor, Robert G. Sachs, knew Chandra during the Second World War, when he shared an office with him at the Aberdeen Proving Grounds in Maryland. Sachs told me about Chandra's tremendous mathematical power, his incredible memory, and how Chandra could repeat verbatim conversations he had or heard several years before. Carl Sagan filled me in on how students at Yerkes perceived him and how, in general, he presented a formidable personality with his scientific accomplishments. Subsequently I had occasion to meet Chandra a few times at conferences. And when he visited Syracuse to give a seminar in 1971, I experienced firsthand his overwhelming personality. During dinner conversations, he brought to life his Cambridge days with stories, anecdotes and reminiscences. Such scientists as Lord Rutherford, Sir Arthur Eddington, P.A.M. Dirac, John von Neumann — all legendary figures for our generation — came alive in Chandra's narrations. It was then that the thought of getting to know Chandra well and learn more about his life, germinated; it became persistent, and led me to approach him during the general relativity conference in Waterloo, Canada, in the summer of 1977, with the idea of writing one or two articles based on conversations with him. To my pleasant surprise, Chandra agreed to talk with me, but he warned me that I was probably wasting my time.

I began taping the first of many conversations with him on 17 December 1977. (I did not know then that soon after his return from Waterloo, Chandra had heart bypass surgery. He gave me no hint of the ordeal he had gone through just a few months before!) These conversations led me to discussions with others who had known him. Chandra also made available to me his extensive files of correspondence with other scientists, with his father and with his friends during his student days. With the surfeit of fascinating material, I could imagine how inspiring his biography would be to many. Extraordinary as it was in terms of accomplishments, it would also be an account of the hardships, struggles and sacrifices one has to make in the process. His story would be the story of students from many lands, who leave behind their native country, the comforts of their cultural surroundings and their loved ones and go in search of learning. It was then that the original idea of one or two articles gave way to writing the biography (*Chandra: A Biography of S. Chandrasekhar*, University of Chicago Press, Chicago, 1991). Chandra was always open to questions about himself and his life, and gave me complete freedom to write whatever I chose to write. It became an overwhelmingly delightful experience to be with him and be transferred most of the time to a different world, a world in which the life of the mind dominated.

Chandra combined with his extraordinary success in his scientific endeavors an equally extraordinary personality characterized by an intensity and fervor for completeness, elegance and, above everything else, gaining a personal, aesthetic perspective in his scientific work. This is best illustrated in his essay "The Series Paintings and the Landscape of General Relativity," in which he draws an analogy between the series paintings of Monet and certain equations of general relativity. In Monet's paintings, the same scene is depicted over and over again under different natural illuminations and seasonal variations. The valley, the trees and the fields, and the haystacks are the same. Superficially, they may appear boring and repetitive. However, the different paintings radiate totally different aesthetic content. In a similar fashion, the seemingly similar equations and solutions in general relativity describe vastly different physics. In concluding that essay, Chandra says he does not know if there has been any scientist who could have said what Monet said on one occasion:

> I would like to paint the way a bird sings.

But we do know of a scientist who spoke like a poet on one occasion:

> The pursuit of science has often been compared to the scaling of mountains, high and not so high. But who amongst us can hope, even in imagination, to scale the Everest and reach its summit when the sky is blue and the air is still, and in the stillness of the air survey the entire Himalayan range in the dazzling white of the snow stretching to infinity? None of us can hope for a comparable vision of nature and the universe around us, but there is nothing mean or lowly in standing in the valley below and waiting for the sun to rise over Kanchanjunga.

With such writings, often filled with parables, quotes from modern and ancient literature with his Ryerson lecture, "Shakespeare, Newton and Beethoven," and his book *Truth and Beauty*, Chandra bridged the gap between what C.P. Snow called the Two Cultures — the culture of the sciences and of the humanities.

Chandra often remembered E.A. Milne, a close friend of his earlier, Cambridge years, and quoted him as saying:

> Posterity, in time, will give us all our true measure and assign to each of us our due and humble place; and in the end it is the judgement of posterity that really matters. He really succeeds who perseveres according to his lights, unaffected by fortune, good or bad. And it is well to remember there is in general no correlation between the judgement of posterity and the judgement of contemporaries.

A year after death is not a true measure of posterity. However, a two-day symposium (December 14–15, 1996) on "Black Holes and Relativistic Stars" in Chandra's honor at the University of Chicago, and this commemorative volume, mark the beginning of that posterity's assignment to bestow on him his due place. Scores of obituaries, reviews and letters have lavished praise on Chandra — the extraordinary scientist, scholar, man of letters, humanist, rationalist, one whose life stands out for its singular dedication to science and for practicing its precepts and living up to its values to the closest possible limit in one's life. Posterity will certainly take all this into account in according him his place.

Finally, I would like to thank all the contributors for their articles illuminating different dimensions of Chandra. I have special thanks for Lalitha for her article as well as for providing this volume with pictures and photographs.

I have received invaluable help in organizing and editing this volume from Annamaria Zajec. I also would like to thank the Editor and his colleagues at World Scientific for the wonderful job they have done.

"My Everlasting Flame"

Lalitha Chandrasekhar

Can one's determination to live to finish one's undertaking conquer over what fate has determined for you? I have a feeling that Chandra had that will and succeeded, although he had the constant fear that he was going to die before me. "Chandra, we can't be sure who will die first. It may well be that I will die before you." The question was raised but we did not know the answer then. The undertaking was his *Principia.* The steel of his determination won over the threat of Fate, as we shall see later. In this indomitable conflict Fate had to accept defeat for a while!

We had planned to leave for Wales on Saturday, the 26th of August. Chandra had written to Bernard Schutz at Cardiff for travel brochures to plan our trip. He had also intended to discuss with him his last paper in the series he had written with Valeria Ferrari on the non-radial oscillations of stars. A week before he died Chandra said to me, "I don't see why I have to discuss the paper with Bernard Schutz since it seems so right." Chandra was extremely enthusiastic about the way the problem was resolving itself. It was all so clear in his mind but he had not described it in detail on paper before he died. I collected his manuscripts at home and his office at the university and e-mailed them to Valeria. She worked on the problem as best as possible but could not discover what Chandra had in his mind. I urged her to look into the matter again. It is my sincere hope that in the course of time others besides Valeria will study the problem. I had a feeling that Chandra was on the verge of something very deep and profound which would throw a new outlook on things. If he had lived he would have succeeded in explaining his ideas to Valeria. This had happened a couple of times before with some of the earlier papers in the series.

There is no question that one of the strongest of our memories of India was its music. Chandra loved to hear me sing. In those days when Chandra used to drive every week from Williams Bay to Chicago to give lectures and also to attend to the *Journal* work, it used to be my habit to sing to him during our long drive back to Williams Bay. This very good habit of mine slackened somewhat after we moved permanently to Chicago. But the interest returned fortunately, and I would say I sang to him very often during the many months before he died. A week before he died, I sang a song to him about Krishna lifting the Gowardhana mountain to cut off the sunlight during the great war of the *Mahabharata*. "Won't you sing it again?" he asked. "No, Chandra. I have another song I want to sing to you now; but I will sing it again later." But that "later" did not happen! The day before he died I had planned to sing still another song to him that I had heard years ago at a concert and had never learned to sing it before! Somehow it came back to me and it was beautiful. It was about Ganapati, son of Shiva. Everyone loved Ganapati, but he was also a scholar, and transcribed the *Mahabharata* when Vyasa dictated the epic. "Shall I sing it to you, Chandra?" "No, Lalitha, I am not feeling well. Some other time," he replied. That "some other time" did not come around since Chandra died the next day. He left soon after breakfast to see Dr. Kirsner at the hospital. He had a massive heart attack on reaching the hospital and died just four hours after he left home.

It was a shattering experience for me. How could it happen so suddenly and without warning! I consoled myself by saying, "He didn't suffer!" But we had hoped to be together some more years and that was not to be.

Chandra's body was cremated and I sprinkled his ashes on every spot in the University Campus that he had been to: the place around the Moore statue, the grounds of the Court Theatre, the Laboratory for Astrophyscis and Space Research, the Crerar Library, the Administration Building, Mandel Hall, the Regenstein Library, and the Seminary Co-op Bookstore, in that order. I had requested Mr. Sonnenschein, the President of the University and Mr. Kleinbard, the Vice-President of the University News and Community Affairs, to accompany me when I sprinkled Chandra's ashes. They were very kind in accepting to do so. And finally Mr. Kleinbard drove me to the Promontory point jutting into Lake Michigan, where I discarded the last of Chandra's ashes to be churned in time to the Atlantic Ocean and from there to all the oceans of the world.

Chandra did not want a memorial service after his death. The idea that I might have a reunion of Chandra's friends and colleagues on an informal basis occurred to me when several of them expressed the wish to have an occasion when they could express what they felt for Chandra. About seventy-five of them from the Physics, Chemistry, Mathematics, and Astronomy departments of the University of Chicago and some friends from outside Chicago gathered together at Swift Commons on October 18, 1995, about two months after Chandra's death. I thanked those who had accepted my invitation to come to the reunion and described in brief how our life in this country had started with Chandra accepting a position at the Yerkes Observatory at Williams Bay, Wisconsin. This institution was a part of the University of Chicago, and Chandra's decision to join it in preference to the Harvard College Observatory was, I am sure, because of the magnetic personality of President Robert Maynard Hutchins of the University of Chicago and also because of the young and enthusiastic faculty that Dr. Otto Struve, Director of the Observatory, had gathered together at Yerkes.

He entered his new appointment at Yerkes with enthusiasm, did a fair amount of teaching, often bearing upon his own research since he felt very strongly that teaching was the best way to clarify one's thoughts while doing research. Apart from teaching, Chandra had also accepted the task of running the Yerkes Library and was in addition put in charge of running the weekly colloquia. So far the colloquia had been on astronomical subjects. Chandra felt that astronomy needed the infusion of physics to make it expand in an astrophysical sense. He persuaded the physicists in Chicago to come to Yerkes to give their colloquia. He also persuaded the astronomers to go to the Chicago campus to give their colloquia in order to influence the physicists. In the same way chemistry and geophysics were also drawn into the picture. Chandra's vision grew wider still and he soon felt that a scientist's life would be the richer if it included among the various influences art and music in addition. This culminated in the Ryerson Lecture he gave in 1975 on "Shakespeare, Newton and Beethoven or Patterns of Creativity".

Going back to the colloquia he ran at Yerkes, he used to number them and keep a record of the date and title of each colloquium in a notebook. When the number reached 100, he celebrated the occasion with a cake decorated with that number! He did the same thing at 200 and so on until the number had reached a full 1000. He had a very special cake made for the occasion with this grand and impressive number showing at the center. Twenty years had elapsed and Chandra felt that it was a good time for him to give the job to someone else. A few years after we had moved permanently to Chicago he went back to Yerkes to retrieve the colloquia notebook. It could not be found. Someone had discarded it in a wastebasket! Why anyone would want to do such a thing was beyond our understanding. What a fine record it would have made for us! It was too bad!

Chandra was the managing editor for *The Astrophysical Journal* for nineteen years. But if you add the six years when he was an associate editor, and that effort was in no way minor, it can be said that Chandra was associated with the *Journal* for a quarter of a century. Chandra thought the job was important. If no one else was willing to do it, he had to do it. One day I asked him, "Chandra, for how long are you going to be stuck with this journal? Is it going to be for ever and for ever? It is standing in the way of our having a fuller life together. Should you not think about my needs also?" There was a moment of poignant silence. Chandra was then swept out of that inner world of his into a stark realization that he had neglected me. He was moved to tears and said, "I am sorry, Lalitha. I should have known better. It will take some time for me to make the necessary arrangements so that I can leave the job." He had to find another managing editor. Chandra had taken care of the letters and supplements also, but no one he asked was willing to take on the entire load. Chandra had done all that, so far, in addition to his regular research and teaching.

The job had to be split between Helmut Abt as the new managing editor, and two further editors for the *Letters* and the *Supplements.* In addition Chandra raised the position of Jeanette Burnett, who was his editorial assistant, so far, to that of an assistant production manager. An office for her management was set up at the University of Chicago Press Building. In addition he had to make arrangements for the American Astronomical Society to take charge of the *Journal* and turn over to them the half million dollars he had accumulated in a reserve fund.

Once all this was done we could think of moving permanently to Chicago. Initially we rented an apartment at 5550 Dorchester in order to get a feeling for what life in Chicago would be like. We used to stay there for only one or two nights a week. Then one day while driving past the new apartment building we had seen growing at 4800 Chicago Beach Drive, I said to Chandra, "Let us go in there and see what kind of apartments they have." "You mean now?" "Yes, now." We went in and saw a three-bedroom apartment we liked and signed the lease right away! That was a huge triumph for me. We were ready to escape Yerkes finally. We had been there too long. A new life was beginning and we looked forward to it. Chancellor Hutchins had been asking Chandra for a number of years, "Chandra, when are you going to move to Chicago? I think you should do it soon." We had done it finally! We stayed three years at 4800 before we moved into our present apartment at 5825 Dorchester. This time it was Chandra who made the move since 5825 was closer to the university. He had signed a lease for a three-bedroom apartment. "Why three? I thought they had four-bedroom apartments," I asked. "Why do we need four bedrooms?" Chandra asked, puzzled. "Well, we had a common study at 4800. With four bedrooms I can have my own study," I replied. Chandra made the change and we have been here for almost thirty years.

Chandra would often remark, "Why do I have to continue doing science? A moment should come when I can say 'No more; no more writing papers, no more of this incessant grind drafting the text, entering the equations, the proofs, etc.' and turn to my other interests in music and literature." Nobody believed he would stop and could stop. *The Mathematical Theory of Black Holes* was written and published — a monumental undertaking. That was the period of his collaboration with an extraordinary person, a student, a colleague, and a dear friend: Basilis Xanthopoulos. His assassination by a deranged student in Crete was a shock Chandra could not overcome. *The Mathematical Theory* was followed by a series of papers on colliding waves in which Basilis also collaborated. Then followed the series of papers on non-radial oscillations with Valeria Ferrari. Yes, no one believed Chandra would or could stop! And when the three-hundredth anniversary of the printing of Newton's *Principia* was celebrated and Chandra was asked to give a talk on the occasion here at the University, Chandra's early desire to read the *Principia* which he had set aside came up to the forefront and he took it up enthusiastically. To begin with, he was going to study only a few propositions bearing upon the universal law of gravitation, but as time went on, his interest grew and expanded. He studied the lunar theory, the figure of the earth, the theory of tides, comets, velocity of sound and conic sections. There was excitement when he described the Brachistochrone problem, the initial value problem and the ascent and descent of bodies. He was thorough in verifying every equation, every corollary, and when Newton just gave the result without providing a proof, he would investigate and find out how Newton must have proved it. One of the most exciting moments in his study of the *Principia* came about when we visited the Cambridge University Library archives in Cambridge, England, to look at Newton's papers. Chandra had drawn a diagram as he thought Newton must have drawn it when he proved a particular corollary. What did Chandra find? A diagram identical to the one he had drawn! I was most impressed and said to him, "Chandra, you have succeeded in entering Newton's mind and discovered how his mind worked."

Chandra was so afraid he might not live to finish writing his *Principia* and when he had finished writing it, to see it through the publication by the Oxford University Press. But fortune was kind and it was published and he had finally his own personal copy of the book that he held in his hand and could turn the pages. But the most exciting moment for me was when he called me from his office to say that he had received ten extra copies. "The first of these copies is for you. May I come over now to give it to you?" "Oh, please do. I can't wait to receive it." He came over right away and presented me with my copy. I turned the pages, and was very soon overwhelmed on seeing that the book was dedicated to me! "Chandra, how good of you, it is an honor, but I am not sure if I deserve it." "Of course, you do. You have been with me all through the writing, supporting me, and following my efforts with great interest. When we were at the Oxford University Press to see the galley proofs of the book, I had told everyone, particularly Mike Mansfield, that this particular page should not be in evidence, and you did not notice the absence!" "It is such a beautiful idea, and is all the more meaningful to me. Thank you for making this moment so very precious to me." Chandra had that rare ability to give a special meaning to such moments. Chandra watched me as I turned every page to see the carefully drawn diagrams, the explanatory texts, and the superb printing by the Oxford University Press. "Beautifully done," I told him and remembered Mike Mansfield telling me, "This is the best book we have published." Chandra used to say that he had discovered something new in almost every page of the *Principia*. It was our intention to go through the entire book together, but unfortunately, we were able to go through only about a third of it. August 21 had arrived to take him away.

Chandra used to say that he would probably be remembered more for his work on the *Principia* than for his other scientific contributions. That he would be remembered for his work on the *Principia* is without question true. But I have no doubt that he would be remembered equally for all his scientific work. The interest, use and influence of any of Chandra's work is due to the way in which he approached the subject under investigation: he looked into every detail, found out what had already been done in the subject, what had still to be investigated, and what were some of the errors that had entered into the field and caused confusion in scientific thinking. The problem became particularly serious if a well-known scientist had introduced the error and people out of respect for him failed to notice the error! When every detail in a subject had been carefully looked into, the subject began to reveal new secrets. These were the discoveries Chandra made since the subject lay transparent before him. He would stand back and get a perspective of the subject. I had an input in the way he looked at a subject. I called the arrangement of all the details a formal garden; the errors were the weeds which had to be pulled out.

There is another thing for which Chandra will be remembered at this university. This was put in a nutshell by President Hutchins to me when Chandra and I went over to hear a lecture of his. He had already left the University at that time to head the Center for the Study of Democratic Institutions at Santa Barbara in California. He had come to Chicago to give his lecture. At the end of the lecture we went over to tell him how much we enjoyed hearing his lecture. Mr. Hutchins then held both my hands and said to me, "The best thing I did for the University of Chicago was to appoint your husband to the faculty." Chandra heard him say it and said to me later, "He is exaggerating." A year later we were in Santa Barbara. "Chandra, we should visit Mr. Hutchins." "I don't know if we really should," Chandra replied. "We really should," I insisted. Mr. Hutchins received us graciously and again as we were leaving, he took both my hands and repeated, "The best thing I did for the University of Chicago was to appoint your husband to the faculty." This time there was no denying the sincerity of that statement. He looked every bit of it and it touched our hearts. We were glad we went to see him since he died soon after. Now, why did Mr. Hutchins make this statement to me on two different occasions? There is no question he must have remembered how Dean Gale of the Physics Department had refused to allow Chandra to lecture at the campus. The refusal was blunt: he did not want this black scientist from India to lecture in his department. Hutchins had written a one-line reply to Mr. Struve, the Director of the Yerkes Observatory, who had been in a dilemma, and brought the matter to Hutchins' attention. "Mr. Chandrasekhar shall give his lectures." The lectures were given and many who had heard them have remarked about their mathematical elegance. The full impact of Hutchin's remark to me was that Chandra had paved the way for other non-white members to be appointed to the faculty.

But long before all this Chandra had been lecturing at Yerkes Observatory. News of these remarkable lectures spread and students came from all over the country to hear him, and also do research with him to earn their Ph.D.'s. Fifty-one students in all received their Ph.D.'s with him. A few of these were from overseas. There were also post-docs and scientists who came over to work with him.

I used to attend Chandra's lectures at Yerkes also. Not many people know that I was a student with Chandra in the same Physics department at Presidency College in Madras, India. He was one year senior to me. Some of the classes were common for both of us. I used to sit in the front row. Immediately behind me was Chandra. I knew his presence and he knew mine. In this way a friendship arose, which, in later years, Chandra used to describe in the following words: "There was a girl I knew then. I left for Cambridge, England. After seven years I returned and married her; and we have lived together happily ever after." That "ever" lasted almost sixty years. We were planning to celebrate our sixtieth wedding anniversary in 1996 by going to the Lyric opera in Chicago to see Wagner's *Ring Cycle*, just as we had been to Bayreuth in Germany to celebrate our fiftieth wedding anniversary by seeing the same *Ring Cycle*. The tickets came but Chandra wasn't with me to see the operas. I took others and we remembered.

There was an azalea plant which was given to Chandra by a friend of his. It bloomed for a while and then there were no more flowers. But I watered it anyway and it became a handsome plant with beautiful leaves but no flowers. After some months it put out buds. Chandra and I used to watch these buds hoping they would bloom some day. They didn't. But after he died one by one the buds bloomed. The flowers were larger and brighter in color than those the plant had before. "Chandra, you are blooming for me," I said to myself. I called the plant "My Everlasting Flame."

Chandra's remark about standing in the valley below and waiting for the sun to rise over the Kanchenjunga has often being quoted. That was the last sentence in his essay on the "Pursuit of Science." Now I am standing in the valley looking at the top of Kanchenjunga. I feel that he is there on the top looking at the Universe around him. He has reached the summit and he has found peace.

Subrahmanyan Chandrasekhar: My Anna[1] 1910–1995

Parasu Balakrishnan

Pen in hand, I pause to think how I can render in words a faint impression of the respected and beloved man — brother, scholar, gentleman, and genius — who made our circle special by being in it and is now no more.

We were four brothers and six sisters, living in a spacious bungalow called "Chandra Vilas", two-storeyed and beautiful with two high arches in each storey, framing a front veranda in each storey, with a fair-sized compound around it. It was situated in the elite locality of Mylapore in Madras, at that time home of eminent judges, lawyers, scholars, authors and musicians, with ancient traditions going back to the dawn of history.

All the four brothers were born in Lahore, now in Pakistan. Our father, Chandrasekhar Subrahmanyan Ayyar, was in government service in the Finance Department, subject to transfers all over India, then undivided. There was an interval of about two years between the three elder brothers (I was the third), and of twelve between me and the fourth, Ramnath. In my boyish fancy, I used to think that we were like Rama and his three brothers (in the epic *Ramayana*), and that there was an affinity between Anna and me, just as there was between Rama and Lakshmana, the third brother. Truth to tell, in the early years, I sensed that my two elder brothers, Chandrasekhar and Visvanathan, were closer together. The bond or association between Anna and me developed later, in the years of separation after he left for England (which was on August 1, 1930). This separation had the effect of bringing us together in spirit, making our comradeship abstract, shorn of every disfigurement of reality, lifting it to the ideal, lonely world of communication by letters.

In August or September 1936, during his short visit to India, as we were walking together, the two of us, of an evening on the beach road in Madras, he observed wistfully, "The relation between a man and a woman, I think, can never be perfect as between a man and a man — such as, for example, exists between you and me. We don't demand anything from each other and there is perfect understanding between us. But it is different in the case of a man and a woman. One always demands something from the other. The relationship is never ideal."

Of the young days of eager, striving idealism, before he left for England, I retain a few but vivid memories. The first picture of those years that come before me is our walking together on the Marina of Madras in the evenings, on our return from school in Triplicane, he with quick long strides, I keeping up with him somehow, drinking in the breeze like a horse (I should add, a diminutive one) and receiving his few words with silent assent and admiration. Those words were few indeed — and none in the grand style — but they created around us a regaling atmosphere within that of the breezy sea-shore. He loved the beach of Madras greatly. I remember how, when he had the privilege of enjoying a car-drive along the Marina with Prof. Heisenberg of Germany (a Nobel prize-winner), who at that time visited Madras, Chandrasekhar told his mother exultantly that the professor had been charmed with the beauty of the Madras beach. This little event occurred when Chandrasekhar was a student in the Physics Honours Class — the date precisely was 14 October 1929 — and like some of the little events in great men's lives, I imagine, this personal contact with the celebrated foreign physicist had a great inspiring influence on him at that time. For the name of Heisenberg was often on his lips. "Heisenberg," he told me once, "can be compared only with Einstein." And again, "What a genius! and when so young he has flashed across the scientific world with his meteor-like brilliance!" Chandrasekhar came into contact with another great foreign scientist during his college course; and that was Professor Arnold Sommerfeld of Germany. By no means were these the first influences in Chandrasekhar's life. For he told me once laughingly how, as a schoolboy, he used to go to the beach to be alone, and there prostrate himself devoutly on the sands with the prayer "O God, may I be like Newton!" "What days they were!" he added to me wistfully after a moment.

2

This native wistfulness of his nature became emphasized by his prolonged exile from his country. His mother's death, less than a year after he left home, affected him greatly. But his mind was made of the stern stuff of discipline.

One year after another, he laid them in turn at the anvil of his work. In the midst of his hectic activity, and the stream of his publications, the moods of his pensive mind assert themselves, probably accentuated by his controversy with Eddington, which would have been ruinous but for his phoenix-like quality of rising from his ashes. He writes to me from Cambridge on 26 June 1935:

> It is ages since I heard from you ... For the first time in months I feel home-sick. I shall return surely before June next year — that is, in less than a year! How I look forward to it! I shall be different, and so will the others be. Six years! How long! You see, I can hardly imagine how things are at home. Memory fails, not because I cannot recollect, but because memory recalls what is no more. Whenever I think of home the scene that almost appears before me is mother lying on the easy-chair in front of the house — I do not know why I always recollect her in a red *saree* ... All gone ...

Again he rushes back to work. He is invited as a visiting lecturer by two American universities. He crosses the Atlantic on 30 November 1935 on board the *Brittanic.* Two days before his departure from England, he writes to me:

> I am going to Harvard as a visiting lecturer in Cosmic Physics ... Do please write to me during your Christmas holidays, particularly as I shall appreciate letters from home, when away from my second home at Cambridge — if you understand what I mean — I love Cambridge!

At Harvard, he delivers a course of ten lectures; and from there he writes to me on 23 March 1936:

> I was planning to reply to your earlier letter from the boat. As things stand, it is more than probable that I return home for some weeks during the months of July, August and September ... You might know that I have more or less decided to accept an invitation from the University of Chicago to join the Yerkes Observatory as a research associate. It is a fairly good position, and it will be of great value to me personally to be attached to one of the really great observatories of the world. I visited Chicago and Yerkes about two weeks ago. I had a very enjoyable time at Yerkes. The Yerkes Observatory is situated on the bank of a beautiful lake — William Bay — and behind it are wild woods — a truly inspiring place. Dr. Struve, the Director, was very nice, and the prospects are altogether fair ... So it does seem that we are to work our lives' purposes in different countries ... Perhaps I am selfish. But Science

> has the traditions only of itself, while art is true in only so far as it weaves human forces in the network of nature, especially of one's own country. Perhaps I am too sweeping, but I am having in my mind Turgenev and Tolstoy. To say this however is not to deny the international appeal of really great literature.

This letter gives the origin of his connection with the Chicago University, where he is now a tradition by himself and a legend. He received a similar invitation from Harvard University, of which however he did not avail himself. The letter also is a testimony to the literary strand in his culture. It is to be noted that he mentions Turgenev before Tolstoy.

He leaves Chicago for a while for Cambridge and packs for home for India — after six years — for just a breathing space — for just a marrying space.

He sailed from Genoa on 31 July 1936 and arrived in Bombay on 12 August. He married Lalitha, whom he had known as a student in the Physics Honours Class in Presidency College, on 11 September 1936, and left with her for England on 17 October.

I recall, on two evenings, our early years were born as we walked on the Marina of Madras. He was a recognized scientist, he had shot into the Indian sky like a meteor — shall I say like Professor Heisenberg in the German sky. But I saw walking beside me an earnest eager student, thinking only in terms of the pursuit of knowledge, warmed immediately by the mention of any high endeavour in any sphere, persuading me, without patronising me, to think highly of myself. Truly here is the seed of greatness, I thought.

Although I have mentioned the important event of his marriage, I shall not forgive myself if, in the account of the interlude, I pass over a comparatively trifling incident — his visit to Mrs. Janaki Ammal, the widow of the late Ramanujan. He and I went together to her modest lodgings in one of dark, dingy by-lanes of Triplicane. Later, she called at his house, and he introduced her to his sisters and even succeeded in making that very shy lady feel at home. He told her how the great professors across the seas revered the memory of her late husband as that of a *guru*, a master. "The other scientists here," he told her, "are worth only the dust on his feet." He informed her that one of the professors in England was writing a book on Ramanujan, and to illustrate it a good photograph of Ramanujan was necessary. The available photographs of him were disappointing. Could she help in the matter? No, it was a pity, she had no photograph of her husband with her ... However ... Yes ... she had with her his passport. Chandrasekhar replied that he would have a look at it.

She was escorted home. He went with her and got the passport from her. The small passport photograph of Ramanujan was a find!

When Chandrasekhar left India, he entrusted the passport to me, asking me to take copies of the photograph to be sent to him, and to return the passport to her. When I had carried out his instructions (keeping however one copy of the photograph with me), he wrote to me from Williams Bay on 18 October 1937:

> Your letter and Ramanujan's photograph. Thanks very much for sending them. I think it extremely fortunate that we arranged to have this — really the first fairly good photograph of Ramanujan. I think it will become the official one as Hardy will probably publish it in his book on Ramanujan.

Years later, on 19 July 1942, he writes to his father:

> I was interested to know that you have read Hardy's *Mathematician's Apology*. I also was enthusiastic over it. Incidentally, have you seen his book on Ramanujan — a book in the Cambridge University Series — a companion to Ramanujan's *Collected Papers*? In Hardy's book on Ramanujan there is a frontispiece photograph of Ramanujan — a photograph which is, in a sense, a discovery — though Balakrishnan had a lot to do with it.

It was from this photograph that about forty years later, on the initiative of Richard Askey and Chandrasekhar, four busts of Ramanujan were made by the sculptor Paul T. Granlund of Gustavus College, Saint Peter, Minnesota, USA. One of them was commissioned through donations secured by Richard Askey from the international community of mathematicians to be presented to Mrs. Ramanujan, which was done in the autumn of 1983. (Mrs. Ramanujan died on 13 April 1994 at the age of ninety-four.) Richard Askey acquired one; Anna and Lalitha two. One Anna kept in his house, and the other he presented to the Raman Institute at Bangalore. Later he presented his to the Royal Society of England.

3

America proved a grateful and congenial home to Chandra. He joined the Yerkes Observatory as a Research Associate in 1937 and was promoted to a full Professorship in 1943. His papers and those of his colleagues and collaborators became a regular feature in *The Astrophysical Journal*, of which he became editor. He was recognized, in the words of an American astronomer, as "one of the leading authorities in the field of galactic dynamics", and he became an integral part of the progress in astronomy in America. Throughout his years in America, Chandra wrote to me about his life in America, about his work and his feelings about home and the people he had left behind. We exchanged letters regularly. Based on our correspondence, I wrote an article on him more than forty years ago by way of paying him a personal tribute at the time he was elected to the Royal Society of London. (The article appeared in *Triveni Quarterly* of Madras in June 1945.)

To continue here, however, I may perhaps indicate my association with him through the years in the second world of books. In his Cambridge years (1933–34) he presented to me a copy of J.W.N. Sullivan's *Beethoven* and a copy of Sir James Jean's *Mysterious Universe*. The former I used in my article "From Schiller Through Browning and Beethoven to the *Upanishads*: A Crescendo of Joy", which appeared in the *Literary Quarterly* of Mysore in January 1993. The latter I used in my book, *The Big Bang and Brahma's Day*, which came out in 1995.

On his brief visit to India from Cambridge in 1936 he made a present to me of a set of fifteen volumes of Chekhov — comprising thirteen of stories and two of plays — translated by Constance Garnett. Apropos of this splendid and sumptuous gift, I may mention that I wrote a biography of Chekhov in Tamil which was published in 1947.

All through his life and after his death (I shall explain the latter phrase presently) I received gifts of books from him. I shall mention these with the dates when available. These comprised: Victor Robinson's *Story of Medicine* on the eve of his departure to the USA in 1935 (it was useful to me when writing my "Rambles of a Physician"; Quentin Bell's *Virginia Woolf: A Biography* with an inscription dated 4 August 1973; two volumes of *Pediatric Clinics of North America* on my request when I needed them in connection with writing some medical articles; three books of Oliver Sacks', *The Man Who Mistook His Wife for a Hat*, *Awakenings* and *An Anthropologist on Mars*; Arvid Paulson's *Late Plays of Henrik Ibsen*; *Julie and Other Plays* of August Strindberg; Peter Brook's translation of Jean-Claude Carriere's play *The Mahabharatha*; T.S. Elliot's *Collected Poems* (1909–62) with an inscription dated 25 November 1975. (This was useful to me when I wrote the article "An Indian's View of T.S. Elliot's *Four Quartets*", which appeared in *The American Scholar* of

Washington (issue of winter 1991).)

Chandrasekhar gave also some of his own books, like *Eddington, the Most Distinguished Astrophysicist of His Time* in May 1984 ("And Eddington is an Honorable Man", if I may modify Shakespeare's line "And Brutus is an honorable man" in *Julius Caesar*), and *Truth and Beauty, Aesthetics and Motivations in Science* on 21 December 1987.

I have set aside for separate mention in order to quote the inscriptions that they carried, J.A.B. Van Buitenen's translation of *The Mahabharatha*, Books 2 and 3, and Henri Troyat's *Chekhov*. The inscriptions respectively were as follows: "For Balakrishnan, in grateful remembrance of the fortnight I spent at his house in Bangalore, 1975, S. Chandrasekhar" and "For Balakrishnan and Shyamala in remembrance of the most relaxed and refreshing two weeks in your home, Anna, 1987".

Again he sent me regularly nearly all his published articles, even scientific, although I have no pretensions to understanding them. The more general articles and lectures include his lecture *Shakespeare, Newton and Beethoven: Patterns of Creativity*, delivered in Chicago on 22 April 1975, *Remembrance of Hans Zeisel*, received by me in June 1992, *The Series Paintings of Claude Monet and the Landscapes of General Relativity*, delivered in Pune, India, on 28 December 1992, *Science in Pre- and Post-Independent India* (undated) and his address on *Ramanujan's Bust*, published in Records of The Royal Society, 1995.

On 19 July 1995, a month before he passed away (which was on 21 August 1995) I received a copy of his book *Newton's Principia for the Common Reader*. He had phoned me earlier that he had instructed the publisher (Oxford University Press) to send me a copy. Acknowledging receipt of the book over the phone, I told him, "It is a glorious book, magnificently produced. I am overwhelmed. You have spared for me a copy when I cannot understand a word of it." "No," he replied, "there are some quotations in it, you can understand them." Probably he was referring particularly among others to the covering quotation for the book from Dr. Samuel Johnson, which would explain to the incredulous reader why he had given the name, such as it was, to the book. It was his swan-song. He had told me several times over the phone before that with this book he would put a stop to all his research and writing. There is something appropriate in this being his last work, rounding off his life in terms of his boyhood prayer "O God, may I be like Newton!" — or in terms of the dictum "the last of life for which the first was made".

About three months after he passed away, I received a book — a splendid one, *Rabindranath Tagore*, by Krishna Dutta and Andrew Robinson. It was mailed to me by Seminary Co-operative Bookstore Inc., Chicago with a letter from the Manager dated 14 November 1995:

> Dear Dr. Balakrishnan,
>
> Before Professor Chandrasekhar died he had asked the bookstore to send you a recent biography of Tagore which has only now been published in the U.S. We send it to you now with our compliments, knowing that Chandra wanted you to have it.
>
> Sincerely,
Jack Cella
Manager

The list of books that I have given is long, but I have to add to this that he has gifted to me over the years large sums of money, and that without his help I could not have gone to the USA and Canada to qualify myself further. He would bestow favours as if receiving them.

I find I have yet to mention what perhaps is the crowning item in the list. When he came to India some time after receiving the 1983 Nobel Prize for Physics at Stockholm, he gave me the replica of the Nobel Medal which the Swedish Academy gives its laureates in addition to the original gold medal. He told me wistfully, "I do not know what to do with it. I am giving it to you." I do not know how best to preserve it for posterity. Incidentally I also do not know what to do with the four lovely porcelain plates with exquisite hand-painted flower-and-leaf-and-bird designs (from Smithsonian Institute, if I remember right) that he gave me.

On my side I may mention that I dedicated my first book (*Pon Valayal and Other Stories*, in Tamil, published in 1942) to him with the inscription (in Tamil) "Dedicated to my brother Dr. S. Chandrasekhar, who shines in America to the glory of the motherland"; also my book in English *Ramalinga, Poet and Prophet* (1984) with the inscription "To Subrahmanyan Chandrasekhar, who made our world special by being in it"; and my book in English *The Big Bang and Brahma's Day* (1995) with the inscription "To the memory of S. Chandrasekhar". Through the years I sent him copies of my published books and nearly all my stories, poems and articles as they appeared in periodicals; and to these he always responded graciously.

4

On being awarded the 1983 Nobel Prize for Physics (which the scientific world knows came to him forty years late) he wrote to me on 4 December 1983, in response to a letter of mine:

> First, I must express my sense of being overwhelmed by the "happiness" which all of you seem to have shared in the announcement of the award on October 19.[2] And this fact that the award has given so much happiness to those whom one really cares for and loves is the only aspect which has given me and Lalitha unalloyed pleasure. For the rest, the award, while gratifying, is not one that I sought, or indeed one that I considered relevant to a scientific career. I am afraid that its significance has been very greatly exaggerated, and it distorts the perspective. Thus:
>
> A lady asked me the other day, "The work that has been recognized was done some fifty years ago. What have you been doing since?" I responded, "They also serve who stand and wait."[3] To which she asked, "Have you been waiting for the Nobel Prize all these years?" My answer was, "There was no Nobel Prize in Milton's time!" She had not heard of Milton (apparently). And the moral?
>
> ... We are leaving for Stockholm tomorrow and shall be returning on 15 December. That leaves me hardly three weeks before we go to Zurich to give the Pauli lectures (a commitment made some eight months ago) and to receive a "Tomala" prize of the principal university in Zurich. Remember my story of the bedecked general?[4] As you know, the original plan was to come to India directly from Zurich. But even apart from "constraints" mentioned by you, the idea of undertaking a further long journey is more than I wish to contemplate now. The fact is that I am very tired — physically, mentally and emotionally. It was only with the greatest effort that I could concentrate sufficiently to write out the lecture I am to give in Stockholm next week. I feel totally drained.
>
> To return to other matters: I found your reminiscences of our years before 1930 most tastefully written: truthfully written, yet excellent.[5] It vividly brought back to me those years. I well remember the "I doubt it" incident — yes, it *was* mother. I shall write again either from Stockholm or soon after our return.

5

In connection with his later years I may dwell on one point of our association which brings out an important aspect of his nature, told in the first person singular.

In the Colloquium on Nuclear Policy, Culture and History, held in Chicago in March 1987, in its third session devoted to "An Ethics for Nuclear Policy: The Bhagavad Gita, or Gandhian Non-Violence or the Middle Way?" he said:[6]

> I now wish to comment on what has been said about the *Gita.* I should like to preface my remarks with a personal statement in order that my later remarks will not be misunderstood. I consider myself an atheist.
>
> ...
>
> It so happens that my brother, Dr. S. Balakrishnan (residing in Bangalore), is a Sanskrit scholar. I had an extended correspondence and discussion with him on the *Gita.* ... He said that if one views the *Gita* as the proclamation of standards of conduct by God, then none of the problems that I had encountered will arise. For example, the statement has been made around this table that the *Gita* instructs Arjuna to kill — but, according to my brother, that is not the way it should be read. As we know, the *Gita* starts with Arjuna expressing his moral conflicts in having to kill his elders and teachers and many whom he continues to love. But the fact which is often overlooked (to which my attention was drawn) is that when Arjuna expresses his doubts and misgivings Krishna reveals that the end of the war has been ordained and that his elders, teachers and all he loves are destined to be killed. Indeed what is going to happen *will* happen and does not depend upon what Arjuna does or does not do. Krishna's particular advice to Arjuna derives from the fact that what will happen has already been ordained and that the only choice before him is to play his role as best he can in a part that has been assigned to him. Therefore it is not an instruction to kill or not to kill; it is rather the part he has to play in a sequence of events whose end cannot be altered; and that is why Krishna reveals to him on more than one occasion that the future has already been decided. In other words the *Gita* does not give you the luxury of playing a decisive role in the shaping of events. The choice open to you is to play your own part as well as you can under the circumstance in which you happen to be involved.

Commenting on the discussions on the *Gita*, Professor Milton Singer, editor of the Report, observes:[7]

> In spite of the varied interpretations of the *Gita*'s message there is one underlining interpretation which almost all speakers seemed finally to accept. This interpretation was first suggested by Chandrasekhar when he pointed out that Krishna's advice to Arjuna could not be interpreted

> as advice how to act in the future, because Krishna reveals a future to him in which the battle has already been fought and Arjuna's kinsmen are already dead.

The discussion between Anna and me on the *Gita* and the Report on the Chicago Colloquium edited by Milton Singer (a copy of which Anna sent me) resulted in my book *The Bhagavad Gita and Nuclear Policy*, published by Bharatiya Vidya Bhavan of Bombay in 1993.

6

After the incense-breathing morn and the voyage in strange seas of thought alone, I come now to the last phase of his life. Loneliness had marked him for her own. It was a way of life for him, as he once told Professor Kameshwar Wali. His last phase was sombre.

More than once he told me over the phone, "I sometimes wonder whether all that I did and accomplished in my lifetime was really worth the price." And also:

> I have a strange feeling in regard to my book *Principia*. It does not seem to be mine, whereas my previous books seemed to be mine. Having finished the book, I have only the exalted feeling that I had been in the company of a great mind. The previous books represent to me years of hard work and the attainment of an overall perspective of their themes. The feeling with *Newton's Principia* is different. I don't feel it to be mine. I recall only having been in the company of a supreme intelligence.

Again when he was in Madras in December 1994 during his visit to India which he undertook to see his/our sister Savitri, who was ailing with a terminal illness, he told me that he had vague apprehensions of going into nervous exhaustion, which on two occasions, once some months previously, and the other recently, amounted to distinct forebodings of a breakdown. In response to this I wrote to him in my letter dated 21 January 1995 when he had returned to Chicago:

> It was disquieting to me to know of your state of mind which you mentioned to me. Your sense of foreboding and your sense of events happening outside of you including your work and giving you unrest is a vague yet real disquiet. This state of things falling away from one is deemed desirable by the *Upanishads*, according to the teaching of which

however there should be no accompanying disquiet or restlessness but on the other hand serenity.

Again I wrote to him on 8 July 1994:

I feel good getting phone calls from you ... You seem to be in a troubled spiritual state, and to this I am responding:

To me you seem to belong to the class of men, in the words of Emerson in the essay *Character*, "the individuals of which appear at long intervals, so eminently endowed" that they "seem to be an accumulation of that power we consider". The power that Emerson was considering was, in his words, "character which is nature in its highest form". He describes such persons, borrowing a phrase from Napoleon, as "victory-organized". What appears like a continuation of this enunciation occurs in his essay *Fate*: "When a strong will appears, it usually results from a certain unity of organization, as if the whole energy of body and mind flowed in one direction. All great force is real and elemental ... Where power is shown in will, it must rest on the universal force ... I know not what the word *sublime* means, if it be not the intimations in this infant of a terrific force". Arjuna, in the *Gita*, got this intimation from Krishna ... Similarly you HAD to do your life-long, life-filling and life-propelling, and shall I add, life-denying scientific work ... You have been an eminent and pliant tool of this elemental force. Perhaps here is no room for regret...

Another thing that I want to write to you is in regard to your "strange feeling", as you put it, that all your books, all your hard work, when the books have been written and the work has been done, seem not to be yours, seem to be something extraneous, entities by themselves, separate and different from you. This is a mystic intimation, on the intellectual level, proclaimed by the *Upanishads* which in fact extends this sense of non-cognition even to one's body, senses and mind. (Note that the mind is included in the list.) The *Gita* also teaches that once you have performed your work, you should have no further concern with it and that it belongs to God. I see that after all Hindu blood runs in you.

And then ... came a phone call to me at Bombay, where I was at that time, on 22 August 1995 at 7:30 a.m. (Indian standard time), that Anna had passed away at noon (U.S. time) at the University Hospital, Chicago, on August 21. With pain in the chest he had driven from his house to the hospital himself, parked the car and went to the doctor's office. The doctor immediately recognized that his condition was serious with a heart attack and admitted him to the Intensive Care Unit. Lalitha was telephoned and she had a chance to exchange a few words with him. (O shades of the years 1974–76 and the bypass heart surgery which saved the Swedish Academy and the scientific world from a historical injustice!)

I realize that my foregoing peroration is out of place in its solemn, sombre and sad context. I may only voice for myself the mourning of Heraclitus's friend for Heraclitus (in the words of William Johnson Cory):

> They told me, Heraclitus, they told me you were dead.
> They brought me bitter news to hear and bitter tears to shed.
> I wept as I remembered how often you and I
> Had tired the sun with talking and sent him down the sky.
>
> And now that thou art lying, my dear old Carian guest,
> A handful of grey ashes, long, long ago at rest,
> Still are thy pleasant voices, thy nightingales awake:
> For Death, he taketh all away, but them he cannot take.

Notes

1. *Anna* — The Tamil word for "elder brother", the appellation by which Chandrasekhar was known in our circle.
2. Incidentally, Anna's date of birth is 19 October 1910.
3. Quoted from Milton's *Sonnet on His Blindness* (author).
4. The story is that a general had his whole chest covered by medals and when somebody asked him, "How is it that you have so many medals?" he replied, "The first was a mistake, and the rest simply followed" (author).
5. I have not used these reminiscences in the present article (author).
6. *Nuclear Policy, Culture and History,* Milton Singer, University of Chicago, Center for International Studies, University of Chicago, Chicago, Illinois 60637, USA, 1987, pp. 84–85.
7. Report on the Chicago Colloquium, Editor's preface, p. 9.

S. Chandrasekhar: A Personal Portrait

Abhay Ashtekar

I had the good fortune of knowing Professor Chandrasekhar — known to his students and colleagues simply as "Chandra" — for over two decades. Throughout this period, he was primarily interested in general relativity and, since this is the field in which I specialize, I came to know him quite well. Given the scope of this volume, therefore, I thought it would be best to devote this article to my most vivid impressions and memories of Chandra. This seemed appropriate especially because, while much has been written about his work and the impact it has had on the science of this century, not enough is known about Chandra as a person; to most scientists, he remains a distant legend, shrouded in mysteries. I hope personal accounts such as this will help dispel some of the mystery and provide a fuller portrait of Chandra's rich personality.

I first met Chandra when I arrived at the University of Chicago as a green graduate student in '71. He had just turned sixty. I had done my undergraduate work in India and to me — as to most other Indian students in science — Chandra's stature was god-like. We had heard of the innumerable discoveries he had made whose meaning and scope we understood only in the vaguest terms. But there was a feeling of awe and admiration and a conviction that for a single person to accomplish all this, he had to be superhuman. And so, I was very surprised when I first met him. Yes, he did have that pristine air about him, and yes, everything he did — the way he dressed, the way he sat in seminars, even the hard-backed chairs he chose to sit on — everything had an aura about it that set him apart. One immediately sensed a refined, dignified and austere personality, just of the type one would expect of a legendary figure like him. Yet, when it came to science, there was unexpected openness. He treated us, students in the newly formed relativity group at Chicago, as if we were his colleagues, his equals. He would come to all seminars, including the ones given by students. He would ask us technical questions with genuine interest.

When discussions began, he seemed to become genuinely young, almost one of us. I still remember the smile that would light up his face in the middle of a talk when he heard a beautiful result. Sometimes, when he had cracked a hard problem, something that he found truly satisfying, he would tell us about it. The joy he experienced was so manifest and so contagious! His active interest in the progress made by students and post-docs is illustrated in the following incident. Once, Basilis Xanthopoulos — a student in the group — and I solved a problem on permissible symmetries of isolated systems in general relativity. It was not a particularly difficult problem and it took us only a week. Chandra stayed after the seminar where results were presented, and asked us: "How does it feel to solve a problem so quickly?" He was genuinely interested and spent quite some time with us. I could hardly believe that someone of his stature, who had worked on infinitely more difficult and central problems, would be so interested! At such moments, he did not seem god-like, high above us. He became one of us. There was direct communication.

Chandra was a master storyteller; I have yet to encounter his equal. He had such a fantastic memory for dates and details that, in the anecdotes he recounted, everything became alive. And his anecdotes ranged from incidents that took place in the lofty halls of the Trinity College in Cambridge to his small cabin in the ship he took across the North Sea when he went to Russia. He would recount the events as if they had happened yesterday. We would later shake our heads in astonishment. For, here was Chandra telling us about a storm he encountered during the North Sea passage in 1934, or his interesting meetings with the then President of the University of Chicago in 1946, with such clarity and in such detail that we could not have matched in describing events that took place in our own lives just a year before!

I still vividly recall the first time that I heard him tell a story. The students and post-docs in the relativity group had organized a potluck dinner. Chandra and his wife Lalitha came with a delicious vegetarian casserole. When it came to coffee time, there was some unease about how the event was going to end. Do we just say good-bye and leave? Students had planned the menu well but hadn't thought of anything specific as an after-dinner activity. So, there was some unease. Chandra got up spontaneously and told us some wonderful ghost stories — one told to him by Dirac! They were short, dry and crisp and we all gasped when the punch line came and then laughed. Then other people got up to tell other stories and the evening ended in a relaxed and friendly mood.

It turned out that, although we both worked in general relativity, the specific problems I chose were somewhat removed, both in style and detail, from Chandra's primary interests. Yet, he was always generous with his time for students like me. After I left Chicago, I saw him about once a year and each meeting lasted for several hours. I listened to Chandra with full attention. He had so many anecdotes and ideas and comments on a wide variety of topics from which I could learn so much. He shared his views on the motivations of scientists, on the true and the beautiful in science, on the relation between science and art, on the giants and masters in all creative endeavors and what sets them apart. He recalled his early encounters with Werner Heisenberg and Niels Bohr, his discussions with George Solti and Henri Moore. He spoke for hours on the aesthetic beauty and the rich physical content of general relativity, how the two are intertwined and inseparable. He expressed his joy at discovering a "strangeness in proportion" in the mathematical theory of black holes. It was all so inspiring! His analysis was always razor-sharp, probing deep issues which had often remained entirely unexplored. He applied very high standards, especially to himself. Sometimes his views — for instance on Einstein's attitude towards general relativity — seemed harsh at first. But invariably I realized that he was just being scrupulously honest, demanding the highest from those whose work he admired the most.

At first, what he said outside the immediate domain of science sounded very intriguing and interesting but it did not really sink in. One of these early conversations took place after he had given his famous Ryerson lecture: "Shakespeare, Newton, and Beethoven, or Patterns of Creativity." During this conversation, he was talking about why it is that in the arts and literature, the quality of work improves with age and experience while in science, generally, it does not. He felt that we often do science in isolation, focus narrowly on our immediate goals and that we are not sufficiently broad in our interests and pursuits. He said he thought one would do better science if one read Shakespeare, particularly his penultimate play, *The Tempest.* I had read the play as a teenager and could not see Chandra's point. So, I resolved to go back and re-read it. This I did — once and then again, a second time. I still did not see the point! I could not understand how it would make my science richer. So, the next time we met, I told him what had happened. He merely said: "Oh really?" and smiled.

But then as years passed, I began to understand more. In the middle of doing something quite unrelated — perhaps giving a seminar — I would suddenly recall something that Chandra had said and understand it for the first time. Invariably, I would perceive the depth of his remark and its multilayered meaning. And then, as years passed, slowly I realized how close all this was to what we generally call scriptures. His remarks were "wise" in the *true* meaning of the word. One often hears that the scriptures have a deep meaning and the deeper one looks, the richer and the more multifaceted they seem. I have not had this experience with any of the religious texts. The closest I have come is with Chandra's remarks. Again and again, I thought I understood what he meant, only to discover another, deeper layer later. And there remain a great many things that I still don't understand even "to first order," so to say. I have still to appreciate that remark about *The Tempest*, for example. But I have not given up. One day, it may suddenly become clear!

As years passed by, I looked forward to these meetings with Chandra with ever greater anticipation. After each of these meetings, I felt elated, inspired, ennobled. Sometimes, the conversation would go on till what I thought were late hours, for I knew that he was an early riser. But as the hours passed, Chandra's voice would often become brighter, his eyes shinier. It was just fascinating to listen to him and to watch him. When the last of these long conversations took place, Chandra was deeply involved with Newton's Principia. He had one anecdote after another to tell about Newton, about his own experience with Principia. He wanted to share his views on what sets great science apart from just good science. His passion for exploring ideas, for understanding things deeply, his constant quest, his extremely high standards especially with regard to himself, were all there. It was at the same time very humbling and very inspiring to see Chandra in the role of the *Man on the Ladder* — the photograph he kept in his office, which is reproduced and described in the opening pages of Wali's biography. Chandra had reached such great heights and yet wanted to climb further, to see greater vistas, to understand things more deeply, not because it would elevate him further in the eyes of others, but simply because that was his nature. It was a unique experience, one that I will cherish forever.

Chandra: The Great Guru

Bimla Buti

My association with Chandra started as of a guru (teacher) and a shishya (student). In 1959, Chandra was giving a course in quantum mechanics and I, as a graduate student at the University of Chicago, took this course for credit; this was the beginning of our interaction. Chandra, in fact, gave two courses in quantum mechanics; these were the best courses I had at the University of Chicago. He was an absolutely superb teacher. All of us (the students in his class) used to be spellbound. Then when it came to choosing the thesis advisor, I decided to work with Chandra. This indeed turned out to be the greatest decision of my life. However, before officially accepting me as his graduate student, Chandra had very frankly told me that, because of too many commitments, he would not be able to devote much of his time for me or for that reason for any of his graduate students. None the less I accepted the challenge of working with him.

Chandra gave me references of two papers and asked me to go through them. A week later (Chandra used to visit the Chicago campus only once a week; from Yerkes Observatory, which is about 90 miles from Chicago, he would drive on Thursday morning, stay overnight at the International House and drive back to Yerkes on Friday evening), when I saw Chandra, I told him what was lacking in those two papers and how one could proceed to take care of those shortcomings. Surprisingly, this turned out to be a major part of my Ph.D. thesis.

Chandra and His Style of Training

The University of Chicago had a rule. It required the Ph.D. thesis to be based on papers published by the student under his name only. For this reason, unfortunately, I do not have any joint paper with Chandra. In spite of he not being a coauthor of any publication of his graduate students, he would not let us submit any paper for publication or submit the thesis till he was absolutely satisfied with the detailed calculations. Since he did not have enough time to go through the lengthy calculations, he had devised a novel alternative. He made me give a number (there was no upper limit) of after-dinner seminars, where I had to make a detailed presentation of my thesis work. By the time I finished giving these seminars, I had picked up a great deal of confidence, so much so that throughout my professional life later on, I was never afraid of facing any kind of audience whatsoever.

Chandra, even after his stay of decades in the U.S., used to write his papers and books in English rather than in American English. He would always use a fountain pen rather than a ball-point pen, because he was convinced that his beautiful handwriting would get spoiled with the ball-point. I vividly remember his telling me that the mathematical equations form a part of the text and that they should be properly punctuated.

Chandra had the habit of visiting the library very regularly. On Fridays on his return from the Chicago campus, he would stop at the Yerkes Observatory library to browse through the new journals which had arrived during his absence. He, I believe, succeeded in inculcating this habit in at least some of his students. I very frankly admit that whatever I am today and whatever I have achieved in my professional life is due to his silent and admirable training.

Chandra as a Scientist

Chandra was a great mathematician, a greater physicist and the greatest astrophysicist. His classic paper on "Stochastic Problems in Physics and Astronomy" in *Reviews of Modern Physics* (1943) is a crystal-clear witness to this statement. Any scientific problem that he picked up, he would look into all its possible aspects with full mathematical rigor and depth; this is exactly what many a time resulted in a series of papers on a specific topic. All the physicists, and in particular astrophysicists, undoubtedly know what a prolific writer he was, but what most of the people probably do not know is how much time and effort he used to put into the writing of the research papers (after completion of the scientific calculations) or the voluminous books that he had published. He would not be satisfied unless his manuscript was in perfect shape. Even after decades of writing, he would prepare his manuscripts by writing a number of drafts. If you ever had a chance of calling Chandra, in the evenings or over the weekends, when he was writing his book *Hydrodynamics and Hydromagnetic Instabilities*, you would recall Lalitha (Chandra's wife) answering the phone and telling you that Chandra was busy with his $(n-1)$th draft of the book. Chandra wrote several books. I would have liked very much to say a great deal about these books but, unfortunately, owing to space limitation, I leave that to other authors in this volume. I would prefer to share with you more of my impressions and my close and friendly association with Chandra.

The number of books authored by Chandra is a good indicator of the number of fields in which he had worked. He would work in a particular field for a few years and then switch over to a different field, but before this switch-over, he would write an exhaustive book mostly based on his research papers in that particular field. Chandra was of the opinion that the best way to learn a subject is to teach that subject. He would give a course at the University of Chicago in the field that he decided to switch to. In this connection, I would like to quote him (from his letter of April 18, 1978 to me), "This quarter I am giving a course on relativistic cosmology, a subject I have not learned hitherto and which I am learning simultaneously with the lectures. You may ask, why cosmology? During my first year in Cambridge in 1931, Dirac advised me to turn to cosmology with the statement 'If it were not for quantum mechanics, I would be working in cosmology myself.' My response to him was, 'I would rather not, since I should like to have my feet firmly on the ground.' That was forty-seven years ago. I suppose that even in a literal sense, I need not now wish to have my feet firmly planted on the ground."

He, being an excellent teacher, would like to master the subject before giving a formal course. And in order to do that he would give a number of informal after-dinner lectures at Yerkes Observatory. Even for other lectures, seminars and colloquia, he would put in a lot of time and effort preparing them. The level of his presentation would always be compatible with the level of his audience. A reflection of this, one can see in his letter of November 11, 1981 written to me before his visit to the Physical Research Laboratory, Ahmedabad, where I was working. I quote, "With regard to my technical lectures, I shall probably give some aspects of my work on the Mathematical Theory of Black Holes. All of it is of course included in my book; but my series of papers in the Proceedings of the Royal Society (since 1974) as well as my article in the Hawking–Israel Einstein Centenary volume might give a prospective listener a general flavor of my work. I shall not, however, make my lectures beyond the understanding of a normal graduate student of physics; but here he or she will have to be a good graduate student." Chandra indeed was a master in presenting even a very complex subject in a very lucid style.

He always (even before getting the Nobel Prize) drew a real big audience. In 1971, he was visiting the Physical Research Laboratory in Ahmedabad. When I tried to check with him about the title of his lecture which he was to give at PRL, he said, "You announce the title as abstract as possible so that we don't have a big crowd," but to his utter surprise even the abstract title could not dissuade anybody; he had a huge audience.

Chandra as a Person

Ever since I met Chandra, I always found him immaculately dressed up in a formal black or gray suit. He was rather a formal person. To strangers and very often even to his students, he would give a feeling of being a dry person. Fortunately, I had never found him that way. He certainly believed and led a very disciplined life. Even after six decades of his stay in the U.S., he would never touch alcoholic drinks or smoke and remained a vegetarian. This reminds me of a little incident. Chandra, after a long gap of 11 years, was planning a visit to India in the fall of 1961. One day he asked me if I would like to spend some time at the Princeton Plasma Physics Laboratory while he would be away. I was excited and immediately accepted his offer. He talked to Professor Lyman Spitzer and arranged my visit to Princeton for three months. During my stay at Princeton, Lyman Spitzer invited me for the Thanksgiving dinner. I, being a vegetarian, was a little reluctant to accept this invitation, thinking that I could be embarrassing my host by refusing turkey, the main dish for the Thanksgiving dinner. However, when Lyman realized the dilemma I was in, he remarked, "Look, if I can entertain Chandra, I am sure I can entertain you."

Chandra was an extremely hard-working person. When I started working with him, I was told by my seniors at Yerkes that Chandra worked for 15–16 hours a day and probably expected his students to do the same. I was never quite given this feeling by him, but it was certainly true that he would get very upset if he did not find a particular student whom he wanted to talk to, specially about something exciting he had seen in a current journal or about an exciting idea of his own. In some ways, he was very childlike. He had to share his excitement at a given moment. I remember his secretary, Donna Elbert, telling me that Chandra would call her at any odd hour to ask her to check some numerical calculations (Donna used to do his computations of a desk calculator) to confirm whether what he was expecting was indeed correct. Another incident, which I found very amusing, occurred during Chandra's visit to PRL, Ahmedabad in 1982 as a Vikram Sarabhai Professor. One day, I had taken him for some sight-seeing. On our way back, we had to stop at the railway crossing. Chandra very excitedly got out of the car and walked closer to the railway track to have a closer view of the steam engine. Perhaps he had not seen a steam engine all those years in the U.S.

Chandra had a terrific memory. He could repeat verbatim his conversations with many great scientists even after decades. He was a great storyteller too. Once, after moving to an apartment at Yerkes, I had invited some of my colleagues to an after-dinner party. I had asked Chandra whether he would like to join us at that party in spite of my colleagues warning me that Chandra would never accept the invitation since he was known to be rather unsocial. To my surprise and to a bigger surprise to my colleagues, he accepted with great delight my invitation. He and his wife were almost the first ones to arrive and the last ones to leave the party. Chandra hardly gave anyone else any chance to speak. He entertained all of us with extremely interesting stories, anecdotes and reminiscences of great scientists like Einstein, Eddington, Milne, Dirac, Enrico Fermi, Rutherford and many others. My American colleagues, who otherwise probably would be bored at any party without alcoholic drinks, were kept so pleasantly busy listening to Chandra that they were more than happy with soft drinks and Indian sweets. Next day, jokingly I told Chandra that the students were rather surprised that he had accepted the invitation. In reply, he said, "I do not know how people expect me to accept when they don't even extend the invitation." For some reason, most of the people, including his students, used to be scared of approaching him. This reminds me of another incident. During my visit to Princeton Plasma Physics Laboratory in 1961, one day I was having a chat with Ed Frieman, the then head of the Theory Group; I was taken aback when Ed made the statement "Bimla, even as a graduate student, you talk about Chandra as if he is your friend and look at Russel (Russel Kulsrud), he even today (many years after he left Chicago) starts shivering when you talk of Chandra." I bet Ed was joking. But clearly Chandra had created such impressions. Many times, the scientists at many institutions were extremely keen on having Chandra for a seminar or a colloquium but did not have enough courage to make such requests. At some of these institutions, where I had worked in the U.S. at different times, Chandra did give lectures on my initiative in inviting him. But every time, either before or after the lecture, he would make a remark, "I have come here simply because of my student." This used to be very embarrassing for me. Moreover, I know for sure that some of my senior colleagues did not appreciate this statement from Chandra.

I am really fortunate to have had real close relations with Chandra. I could talk to him frankly on all possible issues, including the possibility of his returning to India after his retirement. Unfortunately, some of my contemporaries, at Yerkes, had misinterpreted my frank discussions with him. Wali in his biography of Chandra quotes Jerry Ostriker, "I come into the library, and there was this small woman (Bimla Buti) pounding on the table and saying, 'You are wrong, Chandra, your are wrong.' " This statement is a bit of an exaggeration. However, I could have told Chandra that I did not agree with him.

I used to correspond with Chandra rather regularly for many years after returning from Chicago. He used to reply to each and every letter of mine even though most of the time he would be brief and to the point. You probably cannot imagine how great I used to feel on getting his encouraging remarks on some of my publications and preprints. In his letter of March 1, 1977 he wrote, "I have just read with great pleasure your letter on 'Nonlinear Schroedinger Equation for Dispersive Media' in the latest issue of *Physical Review Letters.* And since I have recently had occasion to lecture on the more elementary parts of the soliton theory (Ala Kruskal and Fadeev), I can appreciate the significance of your paper. I do congratulate you on this beautiful work." Another example from his letter of December 28, 1972: "It does seem to me that defining a temperature T_{ij} to accord with the energy–momentum tensor P_{ij} is more rational than my original suggestion. And it is gratifying that with your new definition all the integrals come out explicitly." Before I had sent my write-up on "Relativistic Particle Distribution Function in the Presence of a Magnetic Field" to Chandra, he had given me some suggestion as to how to go about handling this problem. Chandra had no hesitation in saying that my idea was better than his way of thinking.

Every time I visited the U.S., I made it a point to visit Chicago to see Chandra, who would graciously spare his morning or afternoon, including lunch, for me. During one of my such visits, he was supposed to be in and out of Chicago, but in order not to disappoint me, he wrote me a very friendly letter. I quote from his letter of February 9, 1981,"Just a line to say that I shall be in residence in Chicago during all of March and all of April. But it is conceivable that I may be gone on some particular days. It would therefore be useful for me to know as to when you might be passing through Chicago so that I can keep that week free." During every visit, besides physics, Chandra would be very keen to know about science-related politics in India. In fact, he used to take a great deal of interest in Indian politics. He used to keep himself abreast of the latest in politics. Quite frankly, although he was living in the U.S., he knew more about what was going on in India than I, living in India, knew. He was a great admirer of Mrs. Indira Gandhi. He used to talk about his meeting with her in 1982 with great enthusiasm.

I can go on and on with many more stories, but I must stop. Before concluding, however, I would like to say once again how fortunate I was in having Chandra first as my guru and later as a very kind and a generous friend. My admiration for him only grew with time. I learned a lot from him and, even though he is no more with us now, I will keep learning from the astounding literature that he has left behind for us. Thanks a million, Chandra!

On Working with Chandra

Donna D. Elbert

When Mr. and Mrs. Chandrasekhar arrived at the Yerkes Observatory in Williams Bay in 1937, they became the first people from India to live there. The population of the village was very small and new residents were conspicuous. However, the Chandrasekhars became a part of the community and remained residents for many years. Since I was nine years old at the time, my remembrance of them is one of their always having lived in Williams Bay until they moved to Chicago.

In September of 1948, I applied for the position of technical assistant to Chandra, to replace Frances Breen (who was leaving to raise a family), and learned that the position was to start on Wednesday, January 12, 1949. After being interviewed, I conferred with Mrs. Breen about the work and about Chandra, while Chandra went to the local high school to check on my school records. Thus, we decided to try to work together and we did for over thirty years! I remember thinking that he was such an old man — at thirty-eight years old.

The work that I was hired to perform consisted mainly of using a desk calculator to numerically solve the mathematical formulae that Chandra produced. The other part of the work consisted of typing manuscripts and correspondence.

The procedure Chandra used for answering his correspondence was to have me roll the typewriter on its stand into his office. He would write letters in longhand and I would type them: In this way I could query any unclear words. Since I was inept at typing, it was an extremely trying time for me. It was years before I could change this routine of answering the correspondence. First Chandra wrote his letters and brought them to my office and finally he dictated them to me. He was considerate in not speaking faster than I could write. If he was interrupted by a telephone call, he would resume his dictation where he left off without repeating himself. The title "Distinguished Service Professor" was used only on letters of recommendation written to someone Chandra did not know. One time when he wrote such a letter, I unthinkingly asked if he wanted to add the "blah, blah." He said, "What? Is that what you think of the title?" I was relieved when he took the remark in good humor by laughing heartily.

The manuscripts that I received for typing were always in perfect order and in a form acceptable for direct submission to the publisher. Chandra wrote his manuscripts on lined, loose-leaf notebook paper. Each page was written and rewritten until the wording was as concise as possible and there were no mistakes. Some pages were written nine or ten times to correct minor errors. Before the development of current copying machines, two carbons were typed and I had to fill in the formulae in both copies. Chandra's writing of the formulae was done with a knowledge of how the typesetter would set them in print. This careful attention to the composition of the manuscript for printing illustrates the meticulous manner in which Chandra attended to every phase of his work. His formulae were arranged to look esthetically pleasing in print. When proofs of a paper arrived, Chandra read through them once and I read through them twice. He recorded my corrections and then read the proofs several times again. If I missed a correction he was very displeased.

When his book *Hydrodynamic and Hydromagnetic Stability* was completed Chandra packaged it for mailing and asked me to accompany him to the Walworth Post Office, where mail was sent out late in the day. I sensed his relief in finishing the book and his wish to take it as far as he could. He was intense with every phase of his publications until they appeared in print.

Chandra's habits of work in his years at the Yerkes Observatory were very regular. He was in his office at 8:00 a.m., went home at noon for lunch, returned at 1:00 p.m. and stayed until 6:00 p.m., went home for dinner, and was back in his office until 10:00 p.m. And he told me that he started his day at 5:00 a.m. in his office at home. Later he worked at his office at home in the evenings. Some mornings he would show me his many pages of formulae worked out the previous night. I always dutifully praised him for the amount of work he had produced without being able to fully appreciate the accomplishment.

A list was kept of the number of papers (and books) that Chandra had published each year. Our year's end ritual was to record the number on a graph. If the number was less than the previous year, there was a discussion as to the reason; a book had been written, a research problem had taken too long, he had extra lectures to prepare, etc. He always had an intense drive to accomplish research and I became caught up in his momentum in regard to the computations.

The editorship of *The Astrophysical Journal* cut into Chandra's life and deprived him of extended trips. Once when I expressed my opinion that he should give it up, he became very angry and said, didn't I think he knew when he should give it up? Clearly, he wanted no advice on the matter. However, after he did give up the editorship, he admitted that he could have and should have resigned earlier and that he was foolish to have continued with it for so long.

The award that Chandra received during my years of working with him that seemed to please him the most was the Bruce Medal. The fact that the American astronomers had recognized him with this award, before such a recognition came from Britain, made him very happy. He came into my office several times in succession and talked about it.

It is interesting to find that Chandra's devotion to his work inspired the same kind of devotion from others who worked with him. Women who worked in the editorial office of *The Astrophysical Journal* were so dedicated as to give up some Saturdays with their families to work at the office to meet a publication deadline. I found my own loyalty more directed to completing my work than direct loyalty to Chandra. We had our disagreements but I did not lose interest in my work. Chandra's tact was that I worked with him and not for him. This produced the greatest results. He did not make me feel inferior and was always ready to explain anything I asked. If something was bothering me he would help me sort out the difficulty even if it made him late for his class. There were times when he was unduly harsh and I once countered this attitude with "Yes, master," which cleared the air and amused him. If at times Chandra showed anger which I felt was unjustified, I stayed home the next day on the pretense of being ill. Upon returning the following day to the office, Chandra would realize that I had been offended and he would apologize. I could count on him "to do the right thing." When he was reluctant to admit he was wrong, I persisted until he would make the admission.

Chandra could show great restraint in his answers to referees' criticisms of papers he had submitted for publication. One referee from England, with whom he corresponded directly, was extremely critical and belligerent; he sounded emotional too. But Chandra wrote his answer to the criticism in a deliberate and logical manner, serious and polite, masking his emotion. During one period, there were two astronomers who made a practice of systematically criticizing Chandra's papers in one field in which he worked, pointing out some trivial or meaningless mistakes. Chandra told me that they really did not understand his research. However, he responded to their criticisms with polite explanations. They seemed to be making a career of these criticisms. I was never too sure how Chandra felt about the remarks by one of his friends in regard to *Radiative Transfer* that he had "choked the cat with cream."

There were times when others attributed motives to Chandra's actions that were simply not true. He was honest and direct; he was willing to express his convictions and make a stand alone. When he was the Acting Chairman of the astronomy department, he asked the University administration to appoint an Associate Chairman. I heard someone, who did not know the facts, say that he had been "put in his place" by this appointment. I did not feel the necessity to defend Chandra's actions and I did not assume any prejudices he may have mentioned toward anyone. He was too involved in serious matters to allow others to bother him very much. He was not a jealous person in regard to the achievements of others, though I have recently learned that one of his colleagues displayed extreme jealousy towards him. Because he was so serious, Chandra seemed aloof at times and some people felt that he was angry with them. A colleague of his sent Chandra a note of apology for having irritated him with a question after his lecture. Chandra showed the note to me and said he was surprised that his colleague apologized to him since he was not irritated.

Chandra enjoyed taking walks along the shore of Geneva Lake and did so sometimes when he had a research problem to solve. He told me that he would be so unaware of his surroundings that when he did notice, he was surprised at the distance he had walked. He could completely block out anything by concentrating on his research. He told me of an instance, I believe in India, when he was to give an invited lecture and he felt very ill. He asked for a glass of water and put his mind on the lecture to overcome the illness. This ability to submerge himself in his science sometimes served as a refuge from the ordinary problems of life.

In 1956 Chandra spent the summer in Los Alamos. I decided to spend six weeks in New York to attend a design school. I was allowed to take the time off. In 1960 I wanted to attend the University of Wisconsin summer school to take a mathematics course. Ever agreeable to assisting anyone wishing to further their education, Chandra agreed to my going and drove me up to Madison to register. He helped me with the course on weekends when I was in the office working for him. I was able to spend four summers in Madison taking art history and calculus courses. One weekend I brought a book with me on *Greek and Roman Architecture* that had impressed me and I thought I could impress Chandra. I was deflated when he saw the author's name and said he had known him in Cambridge. Several years later I was allowed to take time off to finish a BFA in art.

At one period when I was at Yerkes, I took off an hour a week to continue piano lessons. Chandra decided he would take lessons. The teacher, the daughter of an earlier Yerkes astronomer, as a graduate of the Eastman School was very qualified. However, after several lessons Chandra stopped. He said that the teacher had not spent enough time on the basics for him and he assumed that she expected him to learn very fast.

It became Chandra's habit to give me a book to read when I left for vacation. Thus, I was introduced to the Russian authors and others that he had enjoyed. After returning from a trip from Europe, he expressed his renewed enthusiasm for Kafka's *The Trial*, which he had reread and now found a new appreciation for it. I later received a copy of that book. I was pleased that I influenced his reading of *The New Yorker*. He started a subscription after I brought a number of articles from the magazine to him on scientists who were friends of his.

While still at Yerkes and between research projects, Chandra was sadly wondering what he should work on next. Joking with him to cheer him up, I said, "You better find something soon. I don't want to work for a broken-down astronomer." He laughed and said, "How do you know that you aren't working for one already?"

Exploring General Relativity with Chandra

Valeria Ferrari

My collaboration with S. Chandrasekhar started in October of 1983. We had met in Rome after the X International Conference on General Relativity, held in Padova in the summer of 1983, and he had invited me to work with him on some relations existing between the mathematical theory of black holes and exact solutions of Einstein's equations possessing two spacelike Killing vectors.

I arrived in Chicago a few days after he had been awarded the Nobel Prize. I was afraid that the commitments associated with such an important event would prevent Chandra from working with me. But my fears were unwarranted, because he was more interested in the work we were doing than in giving interviews to the press. Our first paper was completed in two weeks.

For me, this first interaction with Chandra was surprising in many respects. Knowing the breadth and wide range of his scientific accomplishments and having listened to his lectures at conferences, I had nurtured the idea that he was very strict and rigorous, a man totally and exclusively dedicated to science, and so overwhelming that it would be difficult for me even to talk to him. But Chandra turned out to be entirely different from my preconceptions. In our work, for example, he never used his authority to impose his view on a subject; we always discussed and confronted our ideas as if we were on the same footing. At the same time he was an extraordinary teacher, and shared with me his knowledge and the secrets of his technical ability.

I had to change my views also about Chandra's personality. In spite of his strict appearance, he was a very warm person, to whom friendship was of great importance. Although I came to know him only during the last twelve years of his life, from many episodes that he narrated to me, I think that this had always been the case. For example, in remembering Eddington, with whom he had had the famous scientific dispute that strongly affected his life and his career, he never expressed feelings of resentment or disrespect. I was surprised to learn that while Eddington attacked Chandra's work in international conferences (he characterized Chandra's theory of the limiting mass for the white dwarfs "a stellar buffoonery"), in private they remained on good terms, joining for tea or for a bicycle ride. Chandra was convinced that Eddington's opposition to his theory was motivated by honest scientific disagreement, and his enormous respect, admiration and affection for him were unharmed by these events. At that time Chandra was in his mid-twenties. Chandra told me that when he used to see Eddington walking the streets of Cambridge with an umbrella under his arm, he thought that this was the picture of a man who had dedicated his life to the pursuit of science and finally had reached a sense of harmony and contentment. Thinking of his own future, he would think that he would also experience a similar sense of harmony, peace and contentment in his old age. "But," he would add, "it hasn't turned out that way." He had a feeling of disappointment because the hope for contentment and a peaceful outlook on life as a result of single-minded pursuit of science had remained unfulfilled. I used to wonder, how could a man like Chandra have this feeling of discontentment about his life? Chandra did not exactly know the reason himself. However, I used to feel a sense of relief in seeing that the excitation for a new result, or the occurrence of a problem difficult to solve, was always able to divert his mind from these sad thoughts.

Working with Chandra was very exciting. He had an extraordinary ability to concentrate and work, and a mathematical craftsmanship that allowed him to easily find the right path to solve extremely complicated problems. Throughout our collaboration, we adopted this "interaction procedure": I visited Chandra in Chicago for one or two months in summer or in mid- winter during the break between semesters, and we used this time to set up the problem we wanted to solve. In this phase we discussed entire days until we reached a common view of the problem and agreed upon a strategy to employ. Then we worked out the equations separately and, when I was back in Rome, we compared the results of our calculations by fax. This way of proceeding was very efficient because, due to the different time zones between Rome and Chicago, I could send him my notes at the end of my day and he would find them in his office a few hours later when he went to work in the morning, and vice versa. In this way, we were really working night and day. When the work was completed, he would visit me in Rome for a week to write the paper. Though he was happy to communicate using faxes, I never convinced him to use computers. "I do not belong to this civilization," he would say with his characteristic smile.

During the last five years Chandra and I developed a new formulation of the theory of perturbations of stars in the framework of general relativity. To some extent, the reason why he felt compelled to re-examine this problem has to be found in his aesthetic sensibility, matured through the study of general relativity over several decades. The theory of black hole perturbations shows that the curvature generated by a point-like mass appears in the equations as a potential barrier extending throughout spacetime. All physical phenomena occurring near a black hole, including its oscillations due to an external perturbation and the consequent emission of gravitational waves, can be studied in terms of scattering of gravitational waves of different frequencies on this potential barrier. If this is true for a black hole, it must be true for a star, which is basically a distribution of matter possessing the same symmetries of a black hole. Chandra's experience suggested that general relativity should allow for this symmetry, and he wanted to explore its consequences. The first confirmation of this idea was the existence of a conservation law that we proved in 1990 for the perturbations of a spacetime with a perfect fluid source appropriate to describe a non-rotating star. In principle, this law allows one to establish general relations between the amplitude of an incoming and a scattered wave, and its existence for perturbed stars indicated that, as for black holes, the theory of non-radial perturbations of stars could be reformulated as a problem in scattering theory. In fact, a few months later we showed that, independent of any assumption made on the equation of state of the perfect fluid composing a star, it is possible to write a set of equations involving only the perturbed metric functions, and that the fluid variables can be expressed in terms of them. This result was in complete harmony with the spirit of general relativity, which states that a gravitating system is essentially described by the geometry of the spacetime it generates.

This approach revealed new interesting phenomena, as for example the resonant behavior of the axial perturbations, and the coupling of axial and polar perturbations induced by the dragging of the inertial frame in slowly rotating stars. Chandra was satisfied about our work only when the new pieces of the theory we were constructing harmoniously joined to the pre-existing background. This aesthetic criterion ("a quest for beauty") was a guide in his pursuit of science. He liked to quote a definition given by Heisenberg which very well resumed his thought: "Beauty is the conformity of the parts with one another and the whole." His theory of perturbations of black holes, the exact theory of colliding waves and the theory of perturbations of stars, to which I had the good fortune to contribute, are beautiful examples of this approach.

In 1992 Chandra undertook the study of Newton's Principia, and prepared his last book, *Newton Principia for the Common Reader*. The last time I saw him in Chicago, a couple of weeks before his sudden death, he expressed his feeling of satisfaction for having dedicated the last years of his life to the study of Newton's magnificent work, for he had been able to penetrate and comprehend the intellectual achievements of a genius who had no parallel in the history of science. I vividly remember his words and his smile. "From any Proposition," he said, "I have learned something that I did not know!"

Learning is what Chandra humbly pursued throughout his life. This is a lesson we should not forget.

A Remembrance

John L. Friedman

Chandra entered general relativity in the early 60's when its role in astronomy was first becoming clear. Eddington had taught the subject in a series of lectures in 1931, shortly after Chandra came to Cambridge as a graduate student. But the "hardly veiled contempt [Chandra] could sense, which physicists like Bohr and others had for the work of Eddington (on fundamental theory) and Milne (on kinematical relativity),"[1] dissuaded him from research in relativity; and he did not take it up again until he could write that "the existence of the Schwarzschild limit has been the subject of much recent discussion in the context of the astronomical discoveries pertaining to the 'quasistellar' radio sources."

Even then, he was hesitant. When he asked his friend Gregor Wentzel about going into general relativity, Wentzel said, "If your efforts do not succeed, does it really matter? Why not pursue what you wish to?" Chandra was not entirely convinced. For a short time, he intended to study the subject with an undivided mind, but an unexpected success by Norman Lebovitz in finding an elegant exact solution to a problem that Chandra thought one could solve only approximately, changed his direction. In the next decade, he devoted most of his efforts to what he called "the resuscitation" of the theory of the classical ellipsoids, in collaboration with Lebovitz and others. Nevertheless, his first key paper showing the existence of a relativistic instability that leads to gravitational collapse was completed and corrected early in 1964, "just in time: Misner, Zapolsky, and Fowler were already on the trail."[1]

Despite the fact that he was still sole editor of *The Astrophysical Journal*, Chandra spent as much time on research as did his most dedicated students. Beginning his work by 5 am, he finished each 13-hour workday late in the evening. As part of his moral instruction to us, Chandra did not hesitate to point out that by the time his colleagues arrived in the morning, he had already put in half as many hours as they would work in a day. He described a visit to Caltech mainly by noting that the physicists had spent several evenings during the week at cocktail parties. How, he asked, could they get anything done if this was the way they lived? If a few supremely talented physicists could afford such lapses, Chandra placed himself (and, of course, us) among that vast majority for whom success in science was a matter of character.

When he was in Oxford for six months in 1972, Chandra took a brief vacation with Lalitha. He told me it was the first they had taken for more than a decade. When he and Lalitha invited my family to Yerkes for a day, he pointed out the house where they had lived before moving to Hyde Park. He had mowed the lawn on weekends, he said, and when they moved, they rented an apartment to avoid the lost research time that several weekend hours devoted to housework had entailed.

Character was not an end in itself, but part of the respect that a scientist must accord the natural universe; and the character that allows one to understand nature includes an appreciation of elegance. When I was at Chicago, another American graduate student came to Chandra with a paper written in an English that was not clean enough and perhaps not clear enough. Chandra, obviously dissatisfied, had no specific complaints. Instead he instructed the student to read Shakespeare's *King Lear* and then rewrite the paper.

Chandra presented himself in a black suit with a white shirt. At times, when it was warm or he was less formal, he dispensed with the suit coat. My wife and I cannot recall his ever wearing anything that was not black, white, or gray. He presented his papers with titles that are similarly uncompromising in their disdain for glamor, and our first joint paper was typical: "On the stability of axisymmetric systems to axisymmetric perturbations in general relativity. I. The equations governing non-stationary, stationary and perturbed equilibrium." Looking at the glibly written review of work on black holes by another senior physicist, Chandra was incensed at the way the man glamorized the work of his students.

At the first of the teas to which my wife, Paula, and I were invited, the graduate students arrived, rode together up the elevator of the Dorchester high-rise, and were greeted, ushered into a pristine apartment, and given tea and cookies. We listened in silence to three or four classical records, rode soundlessly together down the elevator, and left. When arguing physics with his students, however, Chandra discarded his formality. And he was often solicitous of us, bailing me out of jail when I was given a traffic ticket after arriving from New Haven without an Illinois license and without the money to post bond; Chandra, himself, went to Marshall Fields to buy my newborn son, Mack, a present. In later years he entertained the whole family at Yerkes with Lalitha, who supplied him with openings for his trademark tales of astronomers and physicists and who provided all of us with vegetarian lunches.

By 1969, shortly before I began working with Chandra, he had summarized more than forty papers on classical ellipsoids in a Cambridge monograph and had turned wholeheartedly to general relativity. Referring to himself as a graduate student in relativity, he was proud that it was more natural for him to interact and collaborate with students than with colleagues his own age. To bring himself and his students up to date in general relativity, Chandra brought a spectacular array of tutors to the University. The year before I came to Chicago, in 1966–7, Roger Penrose had presented the Newman–Penrose spinor-based formalism that later played a key role in understanding the normal modes and stability of black holes. The next year Kip Thorne taught a quarter of relativistic astrophysics. Our own private summer school followed in 1968: With the clarity and elegance of his lectures serving as his argument, Robert Geroch converted us to a covariant viewpoint and a formalism tied to abstract indices that became a signature of Chicago's younger relativists. Geroch and Brandon Carter taught us about black hole uniqueness; in fact Carter had barely finished his uniqueness theorem before giving his lectures. George Ellis replayed for us his influential Cargese lectures on relativistic cosmology.

The next year we were treated to Andrej Trautman, as meticulous as Chandra in his attention to historical scholarship. Trautman taught us and the rest of the physics community that gauge theories were the physicists' rediscovery of E. Cartan's connections on fiber bundles — the generalization for both communities of the electromagnetic vector potential. The Sciama–Kibble theory of torsion was a similar rediscovery (and extension) of Cartan's generalization of the metric connection. It was a couple of years before he (and then we) learned that the Hopf fibration was the Dirac monopole and that both had been discovered in the same year.

In my last year of graduate work, Chandra and Lalitha were scheduled to spend six months at Oxford, and Chandra asked me to come with him to finish up my thesis work, a collaboration with him on the stability of rapidly rotating "configurations," none of which had, at that time, been observed. Paula, Mack (our six-month-old son), and I traveled to Oxford in time for the great blackout of '72, one of the miners' strikes.

In the darkness of that winter, when Chandra went home to his apartment with Lalitha and I to the row house we rented from the Rev. Gauntlett of Maid Marion Way in Nottingham, we worked by candlelight. It was dim and as damp as England's winters have always been. Although Mack was not happy that we were unable to heat his bottles, he forgave us. (England, however, never forgave the miners that cold, dim winter, and she repaid her unions with two decades of Tory rule.) I might have been feeling a little down myself, tired from our son's cries and straining to check equations in the dark. But when I came in to work, Chandra's meticulous script was as elegant as ever, lengthy error-free art, ink on bond. He smiled with mischievous pleasure that I had also been working by candle. Amid 13th century stone walls, built to sequester from the town a secular clergy that once comprised Oxford, he was obviously proud that we each had again spent a day and an evening showing our devotion. It was, he said, as if we were medieval scribes.

The beautiful hand in which his equations were written mirrored Chandra's understanding of the equations themselves. For most physicists on the mathematical side, equations are viewed abstractly in a way that highlights the properties their expressions share as operators on a Hilbert space, while astrophysicists usually take from mathematics only what is needed for the problem at hand. Chandra, however, fell in neither camp. For his time, Chandra was, to my knowledge, unique in the way he treated the equations of relativistic astrophysics seriously as objects in themselves, their structure clear in the manner he displayed them, their meaning to be found in this structure. That mathematics was the language of nature he never doubted, and he served nature all his life.

Chandra was also unique in the way he combined a deep understanding of classical mathematics, of astrophysics and of the history of science, particularly the history of classical physics and astronomy. Trautman and Roger Penrose were then the physicists to whom Chandra seemed closest in temperament and perspective, while his interests were those of the astrophysical relativists, Kip Thorne and James Bardeen. The understanding that grew from Chandra's history distinguished the problems he worked on, and the unmatched artistry with which he handled his language of equations distinguished their solutions. He was as devoted to science as anyone I have ever met. The character that he admired and personified is expressed in the poem by Rabindranath Tagore that Chandra quoted in his address[2] to the Nobel Foundation:

> Where the mind is without fear and the head is held high;
> Where knowledge is free;
> Where words come out from the depth of truth;
> Where tireless striving stretches its arms towards perfection;
> Where the clear stream of reason has not lost its way into the dreary desert sand of dead habit;
> into that haven of freedom, Let me awake.

Acknowledgments

I am indebted to Norman Lebovitz for supplying me with sets of notes that Chandra had given him. In each, Chandra summarizes and comments on the work he had done over several years preceding the date on which the set of notes was completed. The time spent writing the present article was supported in part by NSF Grant No. PHY 95-07740.

Notes

1. S. Chandrasekhar, "General Relativity," unpublished notes of April 1970.
2. R. Tagore, *Gitanjali*.

Motivations of a Hero of Science

Norman R. Lebovitz

Abstract

Chandrasekhar was deeply interested in goals, motivations and methodology in related issues of creativity not only in science but also in other intellectual domains. I discuss some of these interests, incorporating certain personal recollections and observations.

1. Goals and Motivations

Chandra addressed the issues of motivations and creativity in several lectures and articles, a number of which are collected in the volume *Truth and Beauty*.[1] In a subsequent article published in *Nature*[2] he emphasizes three things about science and scientific attitudes: (1) the wondrous nature of science, (2) the goals scientists strive for and (3) motivations and sources of personal satisfaction. These are interconnected; it is difficult to separate them in this way. Chandra's sense of wonder at and enthusiasm for science informed his opinions about goals and motivations. I shall nevertheless try to confine my remarks to the latter.

1.1. *Goals*

With regard to goals, he says the following in the *Nature* article: "Do we wish, for example, to quantify new knowledge by the extent to which others can share in it and still others can make use of it for human delight or for human welfare?" His answer to this question is negative. There is not the slightest hint in this or his other essays of justifying his scientific research on grounds of application or potential application. His own goals are personal: they are intimately bound up with his own satisfaction in scholarly pursuits. His unequivocal position in this is in agreement with certain other distinguished scientists (e.g., Poincaré, Wigner).[3]

What of awards and celebrity? A famous mathematician, when asked what he wanted of his career, is reported to have replied: "The grudging admiration of a few of my colleagues." Like anyone else, Chandra was interested in the good opinion of certain of his colleagues. But large scale celebrity he specifically rejected, and I believe his own actions were consistent with this rejection.

Very early on the morning of his seventy-third birthday word was received that he was to share the Nobel Prize in physics for that year (1983) with William Fowler of CalTech. I had previously arranged to have lunch with Chandra that noon, along with two visitors to the university, Edward Spiegel and Nigel Weiss. After hearing the news, I thought that the lunch plan was probably canceled, but called Chandra's apartment to try to check. I succeeded in getting through the layer of university public-relations staff and talked to Chandra. He said he wanted to keep to as normal a schedule as possible, and would meet us for lunch as planned if he possibly could. And indeed he did.

The effort to keep this award in perspective should not have surprised anyone who knew Chandra, but I believe his natural tendency in this direction received reinforcement from a similar event that occurred three years earlier. Chandra was teaching a morning course in general relativity, attended by a mix of graduate students and faculty. One of the faculty members who attended regularly was James Cronin. Very early one morning that term, word was received that he had received a share of the Nobel Prize. When the class met that morning, Chandra was extremely gratified to note that Professor Cronin was in his usual place.

I once made to him a more explicit statement of the rewards of doing science. Some years ago, a new science library was constructed on the campus of the University of Chicago. Engraved on one of the front pillars is the motto "*non est mortuus qui scientiam vivificavit*" (he has not died who has given life to knowledge). I mentioned it to Chandra in the course of a conversation in his office. He seemed ill at ease and changed the subject. It was not that he disagreed with the sentiment: he has expressed it himself in some of his essays. It was rather, I believe, that it was indelicate to discuss it explicitly.

Is the recognition of posterity one's only goal and motivation in doing science? Many scientists have included beauty as a motivation, even the only motivation. However, beauty in science, as elsewhere, is in the eye of the beholder. In his essay "Beauty and the Quest for Beauty in Science" Chandra addresses this question with quotes from Poincaré, Heisenberg and others. He seems to like particularly Heisenberg's characterization of beauty as "the proper conformity of the parts to each other and to the whole." This of course leaves a lot of room for personal interpretation. For example, it leaves open the question whether practical utility, or at least some form of applicability, should be viewed as one of these parts. Chandra's preoccupation with the classical ellipsoids may have been in large measure due to the mathematical beauty of the theory, but would he have been equally intrigued with them if there were no imaginable connection with the physical universe? Likewise, he often extolled the general theory of relativity for its beauty, but would he have been equally enthusiastic if it had no potential scientific application? I don't think so.

However, this does not contradict Chandra's statement that he pursued scientific goals for their own sake, following esthetically pleasing ideas wherever they led him. On several occasions he quotes Wordsworth on Newton:

> The marble index of a mind for ever Voyaging through strange seas of Thought, alone.

Chandra maintains that this "lonely voyaging" is characteristic of many scientists and artists, and he often claimed that he had pursued the "lonely by-lanes of science." Indeed, he felt that as a result he had never been part of the astronomical establishment.

A distinguished astronomer once asked him: "Chandra, when are you going to come to grips with the real problems of astronomy?" Chandra replied by referring to John Ruskin's accusatory opening question in his essay on Sir Joshua Reynolds: "Why did not Sir Joshua — or could not — or would not Sir Joshua — paint Madonnas?" Chandra's reply to the distinguished astronomer was that he did not feel up to painting Madonnas. This apparently flippant remark contained a serious kernel of truth. He pursued avenues that seemed interesting to him and along which he believed he could make progress. The latter criterion excluded for him certain fashionable lines of research, even if he might himself agree to their importance for astronomy.

An expression of his goals, motivations and sources of satisfaction appears on the page facing the preface to his book *Ellipsoidal Figures of Equilibrium.* It is the following quotation from Virginia Woolf:

> There is a square; there is an oblong. The players take the square and place it upon the oblong. They place it very accurately; they make

> a perfect dwelling place. Very little is left outside. The structure is now visible; what was inchoate is here stated; we are not so various or so mean; we have made oblongs and stood them upon squares. This is our triumph; this is our consolation.

2. Method

Chandra's most famous scientific contribution was the discovery of the limiting mass of white-dwarf stars. Coming as it did when he was a very young man, quite unexpectedly, it appears as a deduction of remarkable depth and insight into a new and exotic aspect of science. In his subsequent research long and daunting calculations, often uninterrupted by physical commentary, conveying an impression of mathematical abstraction, became a trademark of his papers and books. If theoretical scientists can be divided up into those whose methods are mathematical, abstract and deductive on the one hand, and those whose methods are experimental, concrete and inductive on the other, a superficial glance might suggest that Chandra belongs among the former. In fact, he not only practiced the inductive approach, he was a most self-conscious and articulate proponent of it. It was not that he failed to value deep and unexpected insights; quite the contrary. It was rather that he had very definite views as to how those insights were likely to be attained.

The concrete and inductive approach is perhaps clearest in his work on hydrodynamics, and in the culminating book *Hydrodynamic and Hydromagnetic Stability* (*HHS*). A succession of specific problems is attacked and solved, in a variety of physical settings. The outcome of the succession of problems is an understanding and an insight into the subject area. In some cases this leads to the formulation of a general principle or theorem, and in others the insight must stand without the articulation of a summarizing principle. His work on relativity proceeded in the same way. The "daunting" equations were typically undertaken to prepare for a specific calculation in a specific physical setting. Even the early investigation into the nature of white-dwarf stars was undertaken, not because he knew or expected there would be a spectacular outcome, but as a systematic extension of earlier work of R.F. Fowler.

I recall my introduction to Chandra's experimental side. I was a graduate student at Yerkes Observatory at the time when he was working on *HHS*. We were unsure whether a certain operator was symmetric or not, and had both worked on it theoretically without resolving the question. He decided to do computations on a special case, working out matrix elements numerically with the help of Donna Elbert, who was his assistant at that time. On the bitterly cold winter day he undertook this, I was persuaded by some of the other graduate students to play football on the Observatory lawn. I recall being out of breath, frozen and slightly bloodied when Chandra emerged from the front door of the Observatory waving a sheet of paper, obviously elated. He stopped the game and excitedly showed me the numerical results: a matrix symmetric to six significant figures. Armed with the confidence provided by this "lesson in experimental mathematics," as he called it, we soon provided the theoretical proof of the symmetric character of the operator.

In his later years, Chandra grew enamored of Monet's serial paintings. He devoted a physics-department colloquium to the analogy between Monet's efforts to capture the essence of objects by observing and painting them in different light and from different perspectives, and the corresponding view of science, particularly theoretical science, that tries to uncover truth gradually by means of a series of separate but related efforts.

3. A Personal Remark

My association with Chandra goes back forty years, when I encountered him as the professor in my first-year graduate course in electrodynamics at the University of Chicago. Subsequently I knew him as my graduate advisor, as my collaborator, and as my friend. Memories of him are many, but the story that most often recurs in my memory is one in which he makes no personal appearance. It is the following.

In the summer of 1969 I attended a conference in the Soviet Union. While registering I found myself standing next to a young Soviet scientist from one of the eastern republics. The social situation was such that it was awkward not to exchange a few words. We discovered after some fumbling that we each spoke a similar level of broken German, so we were able to hold a rudimentary conversation. In the course of this conversation I mentioned that I had worked with Chandrasekhar. I vividly remember his reaction. His face lit up. "Chandrasekhar!" he said, warmly, "*Er ist ein Held der Wissenschaft!*" ("He is a hero of science!").

At that time the word "hero" was used more freely in the Soviet Union than in the west, and it would not have occurred to me to use it myself. However, one common meaning in English is "a man admired and venerated for his achievements and noble qualities."[4] It was an apt description.

Notes

1. University of Chicago Press (1987).
2. Vol. 344, No. 6246, 285–286 (1990).
3. Nevertheless, as I indicated above, I am not persuaded that Chandra's motivations (or those of the others, for that matter) are so completely separate from applications as his stated position seems to imply.
4. Oxford English Dictionary, 2nd ed., Oxford University Press (1989).

Chandrasekhar and the End of Time

Roger Penrose

Qualities of greatness

This world has seen some scientists of extraordinary ability — some who are quick and often arrogant, others cautious and possessing genuine humility. Among that small proportion who are of real and rare distinction are the very few who are truly great. It has been my considerable good fortune to have made the acquaintance of some four or five of those that fall into this final category, but only one of them could I claim to have known at all well — Subrahmanyan Chandrasekhar.

It is my impression that individuals of true greatness may possess some defining quality that is immediately recognizable. I certainly do not refer to what is sometimes called the "great man syndrome", which is an arrogance or pomposity that is more a consequence of standing or fame than of any intrinsic distinction. Instead, it is a quality that is specific to that individual, unaffected by fame or by the passage of time; and it is not (necessarily) shared by others who might also be justifiably called "great". For one of the features of true greatness is individuality — some internal driving force which is that individual's own and which serves to distinguish him or her from the rest of humankind. In attempting to come to terms with the inner drives of such a person, one is therefore presented with an essential problem. The individual's very individuality makes it hard for an outsider to identify with those qualities and drives which are themselves the true source of greatness.

In Chandra's case, there is the additional difficulty that the sheer scope and detail of his work, covering so many different areas with almost encyclopaedic thoroughness, make it virtually impossible for a single person — and certainly for myself — to appreciate properly the individual character of his enormous contributions to these various fields. Even within general relativity, which is the area addressed by Chandra that I know best, the range of his contributions takes them well outside the particular domains in which I have any claims to expertise.

My appreciation of him must therefore be from the particular perspective of a particular outsider. Yet, it may be that my own perspective, when combined with those of others, can add something of value to the overall picture of the greatness of his achievements and his remarkable personal qualities.

Chandra's move into general relativity

My acquaintance with Chandra dates back to 1962, when I first encountered him at the Warsaw International Conference on General Relativity and Gravitation. That occasion had a particular significance for Chandra with regard to general relativity, as it marked his entry into the world of general relativists. In fact, he attended that meeting as a "student", as his way to acquaint himself best with the current activity in that subject.

Why did Chandra have such determination, at the age of 51, to break entirely into a new field, demanding the learning of many new concepts and techniques, where much of the vast expertise that he had built up over many decades would have little direct relevance? It would be natural to suppose, and as I would strongly suspect myself, that it was his desire finally to address the profound conundrum that his early work had thrown up, dating back to his calculations in 1930 on the boat from India to England — that white dwarf stars of more than about one and one-half solar masses cannot sustain themselves against gravitational collapse. It seems clear that even at that time, Chandra was basically aware of the awesome implications of this conclusion, namely that the collapse of the star must eventually take it out of the realm of known physics and into an area shrouded in puzzlement and mystery. But he was by nature an extremely cautious individual, as is made manifest in the modest way he stated his conclusion:

> The life-history of a star of small mass must be essentially different from the life-history of a star with large mass. For a star of small mass the natural white-dwarf stage is an initial step towards complete extinction. A star of large mass cannot pass into the white-dwarf stage and one is left speculating on other possibilities.

He was not the sort who would attempt, without due preparation, to make "authoritative" assessments of the likely fate of the material of a body indulging in gravitational collapse. There are, indeed, still many possible loopholes in the arguments which lead to the final conclusion that has now become an accepted implication of present-day theory — that, at least in some cases, the fate of a body in gravitational collapse must be to encounter a space–time singularity, representing, for the constituents of that body, an end to time!

The issue had been at the root of his difficulties with Eddington, when Eddington had so unfairly attacked his work at a meeting of the Royal Astronomical Society in 1935. Eddington, also, was aware of the implications of Chandra's findings, but regarded this as a *reductio ad absurdum* and preferred to move along his own highly speculative route towards a fundamental theory, thereby rejecting the sound reasoning within the accepted tenets of procedure that had characterized what Chandra had achieved. Chandra appears to have been deeply hurt by Eddington's reaction — the reaction of a man whom Chandra had previously so admired and looked up to. In response, Chandra turned his back on Cambridge and on the immediate problems thrown up by the structure of white dwarfs, apparently devoting his attention entirely to other problems. Yet the question of the ultimate fate of a gravitationally collapsing body must have continued to nag at his physical understandings for many intervening years — even while he was engaging in thorough studies of matters pertaining to quite other astrophysical questions. It is almost as though he had made a tactical retreat, circling around and exploring the details of the surrounding terrain — stellar dynamics, radiative transfer, and the stability of various types of astrophysical structures — before he felt ready for an assault on the profound issue that his early work had uncovered.

His assault was carefully prepared, and required many years of study of the intricacies of Einstein's general relativity. Not only did he familiarize himself with the standard mathematical techniques and conceptual notions that had been developed for that subject over many years, but he engaged the assistance of certain relativists, such as Andrzej Trautman (and even myself), who had specialist knowledge of some of the less familiar modern mathematical procedures, to give series of lectures in Chicago to him, his co- workers, and students.

Chandra's first contributions in which he was able to bring general relativity to bear on astrophysical questions showed that there were additional instabilities, beyond those of Newtonian theory, making their mark earlier than had been expected, and leading even more surely to the ultimate situation of a black-hole fate for a collapsing star. He then moved to the study of black holes themselves, and became fascinated by the beauty of these structures — particularly the Kerr geometry that pertains to a stationary rotating black hole, the ultimate configuration of gravitational collapse. He eventually referred to black holes, in the prologue to his epic book on the subject, *The Mathematical Theory of Black Holes*, as "the most perfect macroscopic objects that there are in the universe".

Chandra and the role of aesthetics

His attitude to black holes clearly illustrates another fundamental feature of the forces that governed his scientific endeavours. If I have seemed to indicate that his researches were driven solely by a desire to probe the behaviour of the physical world at its ultimate limits, then I have created a completely wrong impression. I do not doubt that such considerations were indeed among those which drove him forwards, but there were others just as strong. His fascination with black holes gained as much from aesthetics as from a desire to push forward the frontiers of scientific knowledge. In his later years Chandra became quite explicit as to the importance of aesthetic qualities in science and in his own work in particular. He addressed these questions directly in his book *Truth and Beauty: Aesthetics and Motivations in Science.*

He also delved further into an exploration of the relationships between artistic and scientific values in his remarkable essay "The Series Paintings of Claude Monet and the Landscape of General Relativity". In this, he made comparison between a succession of Monet's artistic creations and some of his own researches into the structure of black holes and colliding waves in general relativity. One of the things that is striking about this comparison is that although the aesthetic qualities of Monet's paintings are expressed in an entirely visual way, Chandra's comparisons relate, not so much to the geometrical or physical nature of the space–time structures that arise, but to the particular symmetries and other aesthetic qualities that are exhibited in the equations themselves.

This brings out what must surely be one of Chandra's very special qualities: his profound appreciation of the beauty of mathematical formulae. This appreciation extended into pure mathematics as well as applied, and he had an especial admiration for the work of Srinivasa Ramanujan. (He often expressed to me his delight in the fact that the only known photograph of Ramanujan was one that he had himself retrieved. Ramanujan had served as an important inspiration for Chandra in his early aspirations to become a scientist.) Chandra's wonderful way with mathematical formulae must have been a quality that benefited him also through his earlier work — and provided a thread of continuity throughout his scientific researches in various disparate fields of endeavour. However, this quality is particularly apparent in his work in relativity theory. No doubt he was struck by the fact that the closer his researches took him to fundamental issues in physics — in the analysis of the very nature of space–time — the greater was the mathematical elegance that he encountered in the equations.

This elegance was not only to be found in the equations that arise from the structure of the Kerr geometry for black holes, but also in the colliding gravitational waves that Chandra, with his collaborators, made a great study of in the main final phase of his work in general relativity. The curiously beautiful interplay between the colliding-wave and black-hole solutions of Einstein's equations held an endless fascination for Chandra, and his comparisons with the art of Monet seem to have arisen partly from his investigations in this area.

Chandra's lightning visits

It is perhaps of some interest that I refer to certain curious and, for me, rather daunting episodes that took place during this period of Chandra's work. On three separate occasions he telephoned me at short notice to ask whether he might consult me on certain technical matters. (I do not believe that he specified the subject, but I had written on the geometry of colliding waves in a paper with K. Khan in 1971.) Would he not be satisfied with telephone conversations or written correspondence, I asked him? No — discussions in person are all that would do. And no — he was not interested in giving a seminar; nor would he wish to be involved in conversations with anyone else. He would take up as little of my own time as was necessary for getting whatever ideas might be needed. Basically he would come for one day and then return to Chicago. He gave me the impression that he had become stuck, on a certain point, and believed that conversations with me might enable him to get over his difficulties. No — there would be no point in telling me ahead of time what detailed matter he wished to consult me about. Though I found this flattering, the responsibility involved was decidedly alarming. As seemed most likely, I would be able to say nothing to him that he had not already thought of.

The first occasion had been in 1985, when I happened to be in Houston, Texas. By then, he had become an expert on the subject of colliding gravitational plane waves, and had mastered the intricacies of the relevant equations to a degree that I could never hope to match. If that topic was what he wished to consult me about, my one hope seemed be that some of my remaining geometrical understandings (not all published) might just enable me to indicate something to him of which he was not yet fully aware. However, if my memory serves me reliably, he had told me nothing over the telephone of the topic upon which he required assistance, so it was with some nervous anticipation that I awaited his arrival. As things turned out, the topic was, indeed, colliding waves; but I was alarmed to find that none of my geometrical knowledge concerning these space–times turned out to be of particular relevance to the problem that was causing him difficulties. I remember feeling decidedly fortunate, therefore, and not a little relieved, that in the course of our conversation a point of clarification did occur to me (concerning the ambiguities inherent in the physical interpretation of the energy–momentum tensor in the situations under consideration) and Chandra seemed to feel that this insight was sufficient to justify his brief trip from Chicago to Texas and back.

I seem to recall that Chandra was rather depressed on that occasion, perhaps feeling that his creative life had reached its end. I tried to reassure him of the absurdity of that supposition (and, indeed, he achieved a considerable amount in the ten years that folowed). He was reluctant to meet anyone else in Houston during that visit, but he agreed that Robert Bryant should join us for dinner, finding in him a kindred spirit of genuine sensitivity.

The second occasion was, in its way, even more daunting, since this time (27, 28 May 1986) I was in Oxford, so his trip from Chicago involved two transatlantic journeys briefly separated by less than a day in Oxford. I remember that he was insistent on having a technical discussion even on the Monday afternoon of his arrival, despite his evident tiredness due to the journey. The topic was another aspect of colliding waves, and I remember feeling that our discussion in the afternoon had not been much of a success. I was genuinely fearful that he might well return to Chicago with his whole trip having been a complete waste of time for him. But in our discussion the next morning I began to understand the nature of the situation a little better and good fortune again struck, some of my geometrical experience coming to my aid in the nick of time! This small insight was enough to set Chandra calculating on his return flight, and upon his arrival back in Chicago he had completed all that he needed. Soon after his arrival, he telephoned back his good news.

On the third visit (which may have been in 1988), again to Oxford, I had perhaps felt a little less daunted by Chandra's imminent arrival, feeling that even if he returned to Chicago empty-handed a two-out-of-three success rate might not be considered to be a disaster! I suspect I was getting a little blasé about the whole thing by then. Over-confident, more likely. In any case, on this occasion, I remember feeling that our two relatively brief discussions never really left Chandra fully satisfied by the time he had to return to Chicago, despite the fact that I had been confidently pontificating on some issue which I was trying to persuade him actually represented the resolution of his difficulty. I was never quite sure whether he did feel that this particular trip was worth the trouble and expense that it caused him. He never expressed any regret to me, but he would have been too much of a gentelman to do so in any case. Perhaps the fact that he never undertook such a visit again speaks for itself!

Other meetings in Oxford; Newton

Of course, I saw Chandra on many other occasions, usually in Oxford. But these times he was visiting also for some other purpose, usually with his wife Lalitha. The two of them were extraordinarily close, and each gave the other genuine understanding and support. On each visit my wife Vanessa and I would make a special point of sharing a meal with them, and these occasions would be remembered by both of us for the dignity and high human quality that the two of them bestowed. Chandra's particular interests, in these later years, included the relations between artistic and scientific values, and in the similarities between the underlying driving forces that guided great scientists and artists. I once gave a lecture at Wadham College Oxford, in which I tried to point out some of the artistic qualities that are present in mathematics. Among the slides that I used was one taken from *The Mathematical Theory of Black Holes* where I depicted one of Chandra's equations as an illustration. Chandra was clearly delighted that I had used an equation of his in an attempt to illustrate the artistic values of mathematics, but I fear that he was disappointed that I was never able to supply him with a transcript of the lecture — although I was eventually able to send him copies of the illustrations that I had used.

Chandra's interest in the arts was broad. In addition to visual arts, he had a particular devotion to literature and music. Indeed, in his Ryerson Lecture at the University of Chicago in 1975 he wrote on *Shakespeare, Newton, and Beethoven or Patterns of Creativity.* He was clearly an admirer of both Beethoven and Shakespeare. His especial admiration for Newton — almost adulation — is revealed in what was to be his final work.

This was devoted to a study of Newton's *Principia*, which he had thoroughly studied in his own individual way. Chandra's approach was to examine each proposition but not, at first, look at Newton's own proof. Chandra would then work to supply his own, using all the modern mathematical tools available to him — and only afterwards would he allow himself to examine Newton's own arguments. In this way, Chandra was able to obtain a special appreciation of the great simplicity, elegance, and profundity of Newton's own demonstrations — qualities that continued to astound him as he examined the *Principia* at greater and greater depth. In preparation for his final book, *Newton's Principia for the Common Reader*, Chandra gave a series of lectures in Oxford (following a similar series he gave in Chicago). I had feared that these might turn out to be rather dry, but Dick Dalitz and I were able to organize these and drum up a respectable audience in vacation time. In fact the lectures were totally fascinating, and I deeply regreted having to miss all but two of them owing to a prior commitment to be away in Copenhagen for most of the period. Chandra expressed his desire for comments and criticisms from the audience, but woe betide any questioner who dared to question the infallibility of Newton!

The book was published just a few weeks before Chandra's death, and he seemed well pleased with it. It is certainly an unusual work (fascinating, but not without its flaws); it is one that many scientists will be grateful for — and it is a distinctive and valuable testament to both great men. It represents a quality in Chandra that one may not easily discern from his earlier scientific contributions. As I have remarked before, his particular mathematical strengths lay mainly in his way with equations, and I do not believe that geometrical reasoning came as easily to him. Yet he held a great fascination for the power of geometry as used, in particular, by Newton as master geometer. Chandra maintained to the end of his life an extraordinary ability to learn when he set his mind to it, even in areas where his own sensitivities may not initially have been as strong as with his equations. His delight in Newton's geometry is as evident as it is infectious.

Chandra: human qualities and aspirations

What are the qualities that stand out in my memories of Chandra? That he was a great and prolific scientist, there is no doubt, and a deeply individual original thinker. He was enormously systematic and well organized, and he worked incredibly hard. He was a rigorous and somewhat autocratic taskmaster, but he had a genuine appreciation of quality in others. He was a loyal friend, reliable, and totally honest. He was deeply sensitive, but proud. He was a difficult man to criticize, and on occasion his pride might get the better of him — but he would be scrupulously generous with his critics if he could be found to be in error. He was polite and enormously dignified: a greatly cultured individual with a feeling for what is valuable in humanity wherever it might be found. He respected life in all forms (he was a strict vegetarian) and had deep appreciation of the works of Nature. He particularly valued the arts and took great pleasure in them, perceiving profound links between artistic and scientific values.

How did he view the status of his own scientific contributions in relation to his initial aspirations? One recalls Chandra's distinctive way of working — reminiscent of the great mathematician David Hilbert — whereby (in essence) Chandra would devote different decades of his life to different topics, culminating each with a definitive book, and leaving each topic behind when he embarked on the next. What does one conclude from this? It might seem that these decades must have represented, to him, completed work that would be neatly wrapped up in the final book. Perhaps so; yet I detected a restlessness in him indicative of a dissatisfaction with what he had ever been able to achieve.

I suspect that his work in relativity theory was what brought him closest to the ultimate goals that he was striving for. He must have derived great satisfaction from his study of black holes, but there were always profound questions left open — and the more that were resolved, the more new ones would appear. Moreover, in his black-hole work, it was the vicinity of the horizon that was being studied, and this lay far outside the central region where the matter of the collapsing star would meet its fate. To gain insights into the nature of this region one must study the space–time singularities — where space and time themselves reach their final termination. Chandra's work on colliding plane waves must surely have been directed towards gaining an understanding of these singularities, for they provide specific models where one can examine the generation of singularities explicitly.

It is inevitable that the results of this work must remain inconclusive, despite the power and insights that Chandra and his associates were able to provide. If the problem of the ultimate fate of a collapsing star — or a collapsing universe — remains unresolved, it is no discredit to him. He opened our eyes to this profound and deeply important problem and he made great strides towards resolving it. Quite apart from all his other achievements, that in itself might be thought to be enough for any man.

Some Memories of Chandra

Rafael Sorkin

When Chandra died, the type of scientific scholar he represented disappeared, and I doubt that any of us now living will see his like again. The triumph of commercial values and the control of science by people who have been trained to think and act like business executives makes it impossible. On a personal level, Chandra's death also meant to me the loss of someone with whom I felt a special bond, but writing about that would not be appropriate here; nor will I talk about his scientific achievements as such. Instead I want only to set down a few isolated memories of (and thoughts about) Chandra — both personal and scientific — which can help round out the picture of the man represented in this volume.

Chandra loved music. His family was deeply involved with Indian music (for example, Lalitha sings, and she and Chandra's sister Vidya both play the veena); and living in the West, Chandra followed the music of that tradition as well. For many years, my father's string quartet (the Fine Arts Quartet) played an annual concert series in Chicago in the Art Institute's Goodman Theater, and Chandra and Lalitha would always be there. Another time, I recall driving to Fermilab for a Fine Arts Quartet concert (or perhaps it was a sonata recital by my father and James Tocco) and again Chandra and Lalitha were there. He even told me once that he remembered the Quartet playing in Wigmore Hall in London. (A musical background is not unusual for physicists, but there is another small personal link between Chandra and me which I would never have suspected, had I not read Kamesh Wali's biography. The freethinking spirit in Chandra's family was nourished by the writings of the famous atheist Colonel Ingersoll, whose great-grandson (or grandnephew?) turned out to be my history teacher at New Trier High School!)

Of course, Chandra's aesthetic sense was most manifest in his physics (especially as expressed in his lectures and in discussions). For him, a physical theory was above all a system of equations, and it was important that the equations form a harmonious whole, that the connections among them be clarified and made complete. If unexpected relationships turned up, their origin had to be uncovered. Miracles were not welcome, only clarity and perfection (in the sense of completeness). Of course, aesthetic perfection normally requires unity, whereas one might easily see the opposite in Chandra's systematic abrupt changes of his field of study. Nevertheless, his life's work was much more of a quilt of pieces cut from the same cloth than a collection of disparate creations. What remains basically constant in his work is not the topic under study, but the perspective from which the work is done. The Chandra of his stochastic process review is recognizably the same as the Chandra of his most recent papers on black hole perturbations, for example.

I think the aesthetic concern of Chandra's stemmed from something deeper than just a wish to lend his mathematics a beautiful form. He must have viewed a physical theory as a literal description of reality, so that in clarifying its structure, he was clarifying the structure of nature itself. Beauty for him was almost the same thing as truth, not an aesthetic gloss to be applied once the truth had been found out. Thus, his frequent comparisons between science and art were not just comparisons between two arenas of cultural achievement, but between two different modes of seeking truth.

Chandra once told me that when he wrote a scientific article, he had no particular reader in mind. He just wrote to express for himself the understanding he had gained. Although this could not be literally true, it does illustrate the detachment and scientific independence he cultivated. Given this independence, he could purify his motivations for doing physics, by separating himself from the desire for money, honors and official recognition, thereby freeing himself to pursue the deepening of his perspective. I recall, for example, that on one occasion when the Gravity Foundation prize was mentioned, Chandra remarked that he failed to understand how anyone could willingly subject his or her work to that kind of external judgment, by submitting an essay to the contest. (In the same vein Chandra, according to Kamesh Wali's biography, more than once turned down offers from other universities without so much as mentioning them to his colleagues, let alone soliciting a "counter-offer". Can anyone imagine that happening today in an American university?) In this turning away from the trivial, Chandra was entirely consistent. I never heard him refer disparagingly to a colleague's manner of dress, or any other petty detail (even though he himself was always impeccably dressed, perfectly punctual, polite, etc.). If he criticized anyone, it was always for something of substance.

Perhaps surprisingly for someone who sought such a degree of scientific independence, Chandra devoted many hours to reading about the lives of other scientists and mathematicians. He also loved to regale his friends with colorful anecdotes about colleagues. (These accounts appeared to be unusually accurate. When quoting someone who had addressed him, he would, if necessary, even pronounce his own name in the American manner (with the "a" as in "band" rather than as it should be said, more like the "u" in "chuck"). Perhaps in reading so many scientific biographies, Chandra was seeking clues to the unaccountable behavior of Eddington in attacking his classic white dwarf work without warning. Certainly, Chandra was very critical of physicists who lacked the humility to distinguish reality from their own intuitions and presuppositions. He often expressed the thought that this kind of arrogance was a common pitfall for people who had made major discoveries: they became so overconfident of their own intuitions that they were no longer able to learn from nature, or from other students of it. Chandra himself avoided this fate in two ways: through a severely critical attitude to his own work, and through periodically renewing himself by entering into a new field of study.

To someone who didn't know him well, Chandra must have seemed almost English, given his dress, and more importantly his erudition in European literature and music. However, his Indian formation obviously ran much deeper than the stories he sometimes quoted from the *Mahabharata.* Most obvious was his vegetarianism, which in his and Lalitha's case was not explicable as merely a religious requirement, because in fact he was an atheist. But Chandra's Indian-ness also came out in other ways. I remember him telling Fatma (my spouse) and me about fourteen years ago that he had reached the stage in life where tradition required him to detach himself from the world and go off into the forest to seek wisdom. (We asked him where this would leave Lalitha, but she remarked at once that not only men could go off to become "sadhus".) When we met about a year later, I asked Chandra why he was not in the forest. He answered that he had grown too used to his way of life in Chicago, and hadn't been able to abandon its comforts and familiarity.

Chandra's Indian background also stayed with him in a way that had nothing to do with his inner nature, but had everything to do with the racist nature of European societies. I recall him and Lalitha saying that they had at first found it difficult to find a place to live in Hyde Park for that reason. I myself have wondered whether racism might not have been part of the explanation of Eddington's remarkable behavior, but Chandra seemed sure that it had not. (Not that Chandra is unwilling to attribute such attitudes to a respected colleague, because in another case, he was convinced that an eminent physicist at a prestigious American university had humiliated him out of racism.) At a less consequential level, minor racist incidents were part of the fabric of Chandra's life at the University of Chicago, in the early years before he became a Hyde Park icon. Two such which I remember Chandra relating to me occurred over lunch at U. of C. eating places. In one case, a table-mate, looking in the direction of a Sikh man wearing his turban, turned to Chandra and asked: "Why is that man wearing a towel on his head?" (Of course, the cloth in such a turban bears not the slightest resemblance to that of a towel. When Chandra told me this story, I suggested that an appropriate answer might have been: "His religion requires it, but tell me, why are you wearing that rag around your neck?") On another occasion, a colleague asked Chandra in an embarrassing manner whether he wore traditional dress when he visited India. In that case, Fermi rescued Chandra by breaking in with "Of course he does".

In his later years, Chandra seemed to feel more isolated scientifically than he would have liked. He often complained, for example, that colleagues to whom he spoke about his work listened attentively, but turned their minds to other things as soon as the conversation ended. On the other hand, although he had stopped taking students, he still had young people to work with, primarily Valeria Ferrari and (before his tragic death) Basilos Xanthopoulos. With Valeria he studied the scattering of gravitational waves by stars, and one question he asked himself in that connection was whether one could associate a conserved energy current with a perturbation, so that one could meaningfully discuss solutions of the linearized Einstein equation in terms of the energy flows they entailed. Since the closest I ever came to collaborating with Chandra was around that question, I would like to recount the developments in a little more detail.

Considering perturbations off a stationary stellar background, Chandra and Valeria had succeeded in finding an expression for a conserved current quadratic in the perturbed metric and fluid variables. But (in his typical fashion) Chandra was not satisfied with this, and wanted to understand where the expression came from. He also wanted to know how to generalize it to the gravito-electric case. To that end he asked himself whether he could derive a conserved current by taking the second variation of the "Landau–Lifshitz complex". (One needs the second variation to get a quadratic expression.) Although it seems obvious that an exact conservation law must hold to every order, whence the second variation separately must also be conserved, it turned out that the second variation not only didn't yield his and Valeria's expression, it didn't yield a conserved quantity at all! When Chandra confronted me with this result (reminding me pointedly that over tea one day I had said that I didn't see why it wouldn't work), I realized what the problem was and that some old work I had done with Bernard Schutz probably contained the key to the solution. The problem was that he had set the second variation of the metric to zero, whereas consistency with the field equations to second order requires (in the pure gravity case, for example) that $\delta^2 g_{ab} \neq 0$. But this makes the resulting expression useless, since it then depends linearly on $\delta^2 g_{ab}$ as well as quadratically on δg_{ab}. The solution was to use a current derived directly from the Action principle (i.e. a "Noether current"). In effect this meant using the "Einstein pseudo-tensor", except that the latter needed to be generalized to the electromagnetic case.

I presented the generalization to Chandra; it differed from the expression he had been trying by only a single term, which he described as "so simple (for evaluation) that one could cry" — but again it didn't work. This time I *was* worried, since it seemed to me that it must work, but after some playing around with the equations, Chandra and Valeria found that after all the divergence did vanish when the perturbed field equations were substituted appropriately. The final clarification appeared in print in the form of back to back articles in *Proceedings of the Royal Society* (Volume A 435), which, as I said, was the closest I ever came to a joint publication with Chandra. (In one respect this clarification remains imperfect, since the fluid contribution to Chandra and Valeria's original conserved current has never been derived from a Noether operator for the fluid.)

In thus working with Chandra, I got an inkling of the incredibly high standards he held himself to. When he and Valeria had still not hit on the right substitutions, he gave me some of his calculations to see if I could find anything wrong. What I found was only a sign error at one point, which was not crucial, but whose presence Chandra said had "humiliated" him. I told him that if this were true, then I would be living in a permanent state of humiliation!

In his youthful work on white dwarf stars, Chandra was 30 or 40 years ahead of his time, not only in the sense that it took the astronomers that long to accept his conclusions, but in the sense that it took the physicists even longer to think of astronomy as an interesting field for the application and testing of their microscopic theories. In his last work, on the other hand, Chandra looked back 300 years to one of his heroes, Isaac Newton, and showed us how to read the Principia, translating Newton's prose into modern algebraic notation.

His book on the Principia will be the last of his many works we have to remember him by, but the person himself is gone. No more will we hear his stories of other scientists, recounted with so much relish, no more his slightly wheezy polyphonic laughter, no more his characteristically English way of expressing himself, sprinkled with occurrences of "I mean..." and "Isn't it?". But from this life devoted to science we still have something to cherish in addition to Chandra's scientific legacy as such, we have the memory that his type of scholar is possible, and has existed in our time, even if we cannot expect another such to appear in the world we currently inhabit.

Chandra at Caltech

Saul A. Teukolsky

In the early seventies Chandra spent a sabbatical term at Caltech at the invitation of Kip Thorne. I was a graduate student in Kip's relativity group, and we graduate students were quite intrigued at the prospect of a visit by Chandra. We had heard fascinating stories about him from Kip: how he had found the maximum mass of white dwarfs while on the boat from India; the controversy with Eddington, who did not believe Chandra's result; how Chandra changed fields once a decade, moving into a new area and mowing down all the unsolved problems, then writing a definitive book on the subject; and many more. When Chandra arrived, it was a little hard to reconcile the tank-like image with the gentle, unassuming man we met. Only the twinkle in the eyes and the slight smile hinted that we should not be misled by the placid exterior.

Chandra's purpose in coming to Caltech was to learn about the exciting new developments in general relativity, in particular about black holes. About eight or ten years earlier, he had published his first papers in general relativity, on the oscillations of stars and the stability of white dwarfs. Now black holes had caught his fancy, and Caltech no doubt seemed like an ideal place to get into this new subject. Kip was one of the leading researchers in the field, and much of the research in Kip's group was focused on black holes.

At lunch time, we graduate students would eat in the "Greasy", as the student cafeteria was known. The faculty mostly ate in the Athenaeum, an elegant setting with waiters and white tablecloths. A few mavericks like Feynman ate with the students in the Greasy. Kip was out of town, and none of the Astronomy or Physics faculty came over to talk to Chandra. We noticed that he was sitting alone in his office at lunch-time, so it seemed pretty natural to invite him to join us at the Greasy. We must have made quite a sight: a bunch of rather scruffy students surrounding this very distinguished-looking older scientist, immaculately dressed in suit and tie. Despite the difference in our ages, and the disparity in our contributions to science, we soon felt at ease talking with Chandra. He had wonderful stories to tell us about great scientists of the past he had interacted with. Scientists, like sports fans, engage in hero worship, and are fascinated with the details of great scientists' lives. We learned from Chandra the *real* reason that Eddington led the famous eclipse expedition of 1919 to verify Einstein's prediction of the bending of light by the sun: a deal to keep him out of jail for being a pacifist during World War I. Such tales were the kind of thing that gave life to the dry textbook stuff we were absorbing in large quantities.

And I think also that our infectious high spirits and youthful confidence rubbed off on Chandra. We were too young to really know how little we knew, and with youthful cockiness we believed that all physics problems could be solved if you were smart and worked hard enough.

Chandra had a custom that we call him Chandrasekhar and he call us by our last names until we received our Ph.D. Then we could call him Chandra and he would call us by our first names. At the time, this seemed like a perfectly logical way of bypassing the formalities of titles and other hierarchical distinctions in the academic world: once you were beyond your apprenticeship, signified by passing your Ph.D. exam, you were the equal of anyone.

Chandra's sabbatical at Caltech came soon after he had relinquished the editorship of *The Astrophysical Journal.* Single-handedly, he had raised the quality of the journal until it was the most prestigious place to publish one's paper in astronomy. As I started to receive papers to review, I, like many of my colleagues, soon developed a double standard of refereeing: one for *The Astrophysical Journal*, and one for everything else. Today, when a new issue of *The Astrophysical Journal* arrives, which it seems to do with alarming frequency, I find very few really good articles to read per issue. Occasionally I look up an article from a 1960's issue, and I am always amazed to see how many other classic articles appear alongside it. I understand the obligation of journals published by professional societies to be archival, but still I feel sad for a lost era.

It was quite noticeable how isolated Chandra was at Caltech. Despite the warmth and humor that was so clear to us, none of the eminent astronomers there came to talk to him. Chandra once remarked on this to us, and speculated that it might be because at one time or another he had tangled with each of them in his role as editor of *The Astrophysical Journal.* Authors often harbor long resentments over perceived injustices in getting their work published. With hindsight, however, it is much more likely that his overwhelming scientific reputation, combined with his reserved personality, was quite a deterrent, even to the giant egos of the Caltech Astronomy Department!

My own interaction with Chandra revolved around his desire to work on the black hole perturbation problem. The problem was to figure out how a black hole would interact with its surroundings — electromagnetic fields, other massive objects, and so on. Much of the work was being done with the Newman–Penrose formalism, a mathematical framework introduced originally to study gravitational waves but now being used for black holes. The formalism was strange to people with the standard mathematical background for physics, and there were no textbook accounts of it. I was fortunate to have preprints and references from my predecessor graduate students in Kip's group, and so found myself explaining the formalism to Chandra.

After a few sessions, Chandra had absorbed all he needed from me, and began what turned out to be a major assault on the problem. The final outcome of his labors was the publication a decade later of *The Mathematical Theory of Black Holes*, a 600-page treatise dense with equations. I met Chandra a number of times during that decade of research, and interacted with him by phone and by letter. There were times when the only way to describe his state of mind over his research was to to say he was in agony. He, like all of us who had worked on the gravitational field of rotating black holes, had become entranced by the unexpected symmetries and miraculous interrelationships of the mathematical description of these objects within general relativity. Time and again, a marvelous simplicity appeared when one should have found complications. But the complete solution of the problem, giving all the components of the gravitational field of a rotating black hole interacting with an exterior disturbance, eluded Chandra. He filled pages and pages with equations, on paper turned sideways so that the equations would fit on the page. In that sought-after last step, no miraculous simplicity appeared to save the day. In the end, Chandra completed the calculation by sheer force of will. He had to leave out much of the algebra in the account given in Chapter 9 of *The Mathematical Theory of Black Holes.* At the conclusion of that chapter, Chandra was forced to say,

> "In the event that some reader may wish to undertake a careful scrutiny of the entire development, the author's derivations (in some 600

> legal-size pages and in six additional notebooks) have been deposited in the Joseph Regenstein Library of the University of Chicago."

Early on in this research on black hole perturbations, Chandra once chastised me on the way I had written up a derivation in a paper. At the crucial stage, I had written something like "Using equations (9), (11) and (12), we find ... ", and given the answer. Chandra was annoyed because it had taken him a while to fill in the missing steps and verify the result. He felt that I should have given more details of the derivation. I didn't have the heart to remind him of this when I read the above passage in his book!

After the book was published, Chandra asked me why I had stopped working on black hole perturbations when I did. The work that had caught Chandra's interest had been for my Ph.D. thesis, and I did not publish again on the subject. I could have answered him that a beginning Assistant Professor had to demonstrate breadth of research in order to be promoted, and couldn't afford to be labeled as never having got beyond his thesis. But knowing how in Chandra's career he had always followed his own star, pursuing his research with little regard for what others might think, I didn't feel he would appreciate this answer. So instead I answered him honestly: I had stopped working on the subject because I had solved all the problems in it that I was able to. And besides, wasn't he grateful that there were such wonderful problems left for *him* to solve!

Some Memories of Chandra

Robert M. Wald

I first met Chandra in December, 1972, on a visit to the University of Chicago. I mainly recall being quite surprised by his very youthful appearance (even to the eyes of someone in his twenties), and being pleased by the attention and interest he paid to me when I met with him. (I seem to recall that he even drove me to my campus accommodations after the meeting.) However, the main attraction for me in Chicago was Bob Geroch and the large group of bright students who had assembled around him. Thus, when I was offered a postdoctoral position at Chicago approximately one year later, I accepted with great enthusiasm. Chandra's presence in Chicago — while certainly a positive factor — was not a major factor in my acceptance.

During my first year at Chicago (1974–75), my scientific and personal interactions with Chandra were quite minor. Chandra was still recovering from his heart attack when I arrived, but even after he returned to full activity, we did not have many extensive interchanges. I expected this pattern to continue indefinitely. On the scientific side, my interests were then centered on quantum particle creation near black holes, rather than on the relativistic astrophysics and classical general relativity issues that most interested Chandra at that time. On the personal side, Chandra did not seem to be an easy person to get to know, and, in any case, it would not have been natural to expect much personal interaction between an extremely distinguished scientist in his sixties and a postdoc in his twenties.

I was wrong in my expectation of having only minimal scientific interactions with Chandra. Although we never jointly collaborated on any research projects, we had numerous fruitful scientific interchanges. These interactions stimulated several research projects which I undertook, and I have reason to believe that my reactions and input to his work were of value to him.

However, I was much wronger in my expectation of having only minimal personal interactions with Chandra. An event which played a key role in bringing about a much closer relationship between us occurred during my second year at Chicago, when I was still a postdoc but had already accepted an Assistant Professor position. I was generally very happy with all aspects of the scientific and personal environment at Chicago. However, an incident occurred where — in response to a request I had made — I felt that I had been intentionally misled and/or treated in a very cavalier manner. When I sought either an explanation or apology, I was rebuffed and, in essence, told that if I had any complaints, I should put them in writing. I was quite angry, so I did: That night I wrote a memo outlining what had occurred and making the case that I had been treated improperly. Since Chandra and another scientist were tangentially involved in this incident, I decided to give copies of this memo to them. Very early the next morning, I dropped off the memo and its copies in the appropriate mailboxes, and went to my office.

Ten minutes later, the phone rang. It was Chandra, who said he had just read my memo and would like to talk to me about it. He offered to come to my office right then, but I managed to insist on going to his. Since Chandra clearly had called me immediately after reading the memo — and he could not even have had time to have read it more than once — I was sure that his main reaction would be one of irritation. Indeed, I fully expected to receive a lecture from him and perhaps even a scolding. After all, I knew that even if the grievances in my memo were completely valid, they were entirely trivial compared with the many injustices he had suffered during his career — and he didn't go around writing memos about them! Worse yet, he might even think that I was partially blaming him for what had happened, and, perhaps, he might react by challenging my interpretation of the events. Thus, I walked into his office prepared to give him a detailed account of precisely what had been said and by whom, and, thereby, to make the case that the wrongs done to me fully justified my writing the memo.

I was completely unprepared for what happened next. As soon as Chandra began speaking, it became clear that — at least for the purposes of our discussion — he was taking it for granted that everything I had said in the memo was accurate. He had only one concern: Was it really in my own best interest to write the memo? It was apparent that Chandra thought it was not, but he wanted me to draw my own conclusions about it — not to lecture me about it. He made a series of carefully worded, oblique statements and questions, all directed toward getting me to think about what good and harm the memo might do. In doing so, he was extremely careful not to give me any unsolicited advice. Indeed, I recall that at one point he started to say to me "you should..." — but then he quickly caught himself, said something like "of course, you have the right to do anything you wish...", and he then began again on a completely different tack. I do not have Chandra's ability to recall the precise details of conversations which occurred long ago, so I cannot report here the exact words Chandra used, but they were remarkably effective: Within a very short period of time, I came to the conclusion that it was a mistake to have written the memo. Since it was still quite early in the morning (before anyone else involved had arrived), I then simply removed the memo from the other mailboxes.

Looking back on this incident twenty years later, I still find it remarkable how perceptively Chandra was able to understand my feelings, how concerned he was for my welfare, and how instinctively he knew exactly the right thing to do in that situation. After that incident, I had a strong feeling of trust for Chandra that never diminished with time. This trust was reciprocated, and, over the years, grew into a close friendship.

I have many colleagues from whom I have learned more physics and mathematics than I have from Chandra. But no one has had a greater impact than Chandra on my attitude towards the pursuit of science itself. What I learned from Chandra in this regard was taught by his example and probably cannot be transmitted successfully by any other means. Nevertheless, I shall endeavor to express some of the things I learned from him here, as a means of partially acknowledging my debt to him.

As a scientist, one's work is continually judged by others — who usually do not have a full understanding of it — via job offers, conference invitations, grant renewals, etc. There are significant dangers posed by these judgments: If one's work is not sufficiently recognized or appreciated, it is easy to feel hurt, to become distracted by feelings of bitterness or jealousy, and to make scientifically unproductive efforts to promote one's own prior work. On the other hand, if one receives wide recognition — particularly at an early stage in one's career — it is easy to start resting on one's laurels and, eventually, to become more concerned with enjoying and maintaining one's place in science rather than with continuing to do good science.

Chandra was not any more immune than other scientists to feelings of hurt if he felt that his work — particularly, any current research — was not properly appreciated. What distinguished Chandra from most others was that he would never allow these feelings of hurt to be used as an excuse for not devoting his full efforts to continuing his pursuit of new scientific research. Similarly, although by the time I first met him he had received the highest acclaim for his past achievements, he never allowed himself to rest on his laurels.

When it comes down to it, it is one's own judgments — not the judgments of others — which count. One should choose to work on a problem because of one's inner conviction that the problem is interesting and important, not because of any declarations of interest and importance that may have been made by others. I don't know of anything that irritated Chandra more than to have a seminar speaker take for granted that the problem he was working on was important because others were working on the same or similar problems, and then, under Chandra's questioning, be unable to give a coherent, personal justification for doing the work. Once one has chosen to work on a problem, it is crucial that one set one's own standards — and set them at the highest possible level — for the quality and completeness of the work. It is of little relevance if the work meets some minimal criterion of publishability and/or represents an improvement over what was previously known. What is of relevance is whether the work meets one's own standards and is the best one is capable of. I know of no one who was as thoroughly (and, in many ways, brutally) self-critical as Chandra in assessing the quality of his own work. Finally, when the work has been completed to one's satisfaction, do not bask in the glory of the achievements and certainly do not wait for the accolades of others; rather, begin new research on another topic.

A great deal of self-sacrifice is required to approach as closely as Chandra did to these ideals in the pursuit of scientific knowledge. I am not willing to make as many sacrifices as Chandra did. However, I am certain that Chandra's example has helped me and many other scientists to approach much closer to these ideals than we would have otherwise.

Photograph in remembrance of a talk given at Presidency College, Madras, 1936.

Chandra and Kameshwar C. Wali.

Valeria Ferrari and Chandra at work.

Chandra with his brothers and sisters in Madras, 1936.

Chandra, 1938.

Chandra on the grounds of Yerkes Observatory, 1950.

Left to right: *Donna Elbert, Lillian Neff and Chandra, 1950.*

Chandra and Abhay Ashtekar, 1987.

Chandra, Madras, 1951.
(Courtesy T. S. Sadasivan.)

Chandra, 1942.

Chandra and Lalitha in Williams Bay, c. 1940.

Memories of an Uncle

Radha Hegde

About ten years ago, I accompanied my mother Savitri, one of Chandra's younger sisters, to Chicago to visit Chandra. She had come from India. Chandra was waiting at the airport to receive us, but having arrived early, we found him engrossed in a book — a Japanese novel as I recall. That image of Chandra, of being in the here and now but removed in his own quiet introspective world, is the image of Chandra that many of us carry. I recall the emotional intensity of that quiet moment when years of distance melted away between my mother and her brother.

My mother and I spent a few days with Lalitha and Chandra in Chicago. My uncle reconnected with my mother and many stories were exchanged. I recall our visit to the Art Institute to see an exhibition of paintings from the Vatican, a visit to his favorite bookstore, and then a walk around the University campus. Chandra's guided tour of the University of Chicago campus was quite an experience. His face would light up with excitement as he recounted stories in his exuberant style. To my mother, meeting her older brother was always filled with awe and complete admiration.

Chandra was extremely close to my mother. For over 50 years, they had corresponded regularly. They maintained a closeness and an intense bond in spite of the cultural worlds that separated them. The times they met in India or in Chicago were few and far between. My mother passed away in December 1994, just a week before Chandra was to visit Madras, specially to be with her. I called him from Madras to give the sad news. I still remember the tremor of grief and disbelief in his voice rapidly replaced by a wrenching silence. Distraught as he was, he told my sisters and me that he wanted to honor Savitri's last invitation to visit her in India. He knew that his ailing sister had prepared every detail of his visit in order to make it a memorable experience. So my uncle Chandra journeyed, in what would be his last trip to

India, to visit his sister's home for the last time.

He spent a week with my two sisters, Sudha and Rajini, and me, consoling us in our time of sorrow. He spoke to us of many sides of our mother. The past would be brought to life with his soft tones and we would be transported to another time, another space.

"Your mother, when she was a little girl, used to like to poke things around in the garden. And one day, she touched a soft sticky thing," said Chandra and broke into peals of laughter. "It was a frog and do you know that I told her, now that she had touched a frog, she would become one. I really did upset her very much." The past and present merged and blended perfectly in the childhood stories that he recalled and related in such a delicate manner.

At other times during this visit, Chandra, Ayya Mama, as we called him, would be lost in melancholic reverie. With a great deal of hesitation, he asked us, "Can I sit down at Savitri's desk?" He sat at my mother's desk in complete silence. We showed him the letters he had written to her over the years and across the distance, letters that my mother had filed ever so carefully. Chandra was visibly overcome by the way the letters had been preserved by his sister, starting from the forties down to the nineties. Here were letters, reminiscences, even notes about a trip across the United States that she had taken with Chandra and Lalitha in the fifties. "It feels like a betrayal," Chandra softly mentioned to us, an encroachment to tread on the space of one who is no longer present amidst us. And then I remember him saying that the most difficult thing was to carry on the day to day routine of living after the death of someone you love.

"I have missed so much," he softly ruminated. The fact of having lived between cultures and the loneliness that defines that condition surfaced repeatedly almost as a motif in his conversations. This theme was one that Chandra had often discussed with my mother. In a letter (February 16, 1993) my mother wrote: "This New Year opened with your long-awaited visit home. So much anticipated ... yet these meetings somehow come like a flash and disappear like lightning. Sudha and I were so happy that you could find time to attend her Bharathanatyam concert, that I could listen to Mahler's music with you — a rare and rich experience for me." It seems in one of his conversations with my mother, Chandra had reflected and asked whether he had done justice to himself and others by his single-minded dedication to science. My mother, who used to write a great deal, wrote a poem which she called her dreamy answer and sent it to Chandra — I quote my mother's rendition of the question Chandra posed because I think it captures not only their closeness but the poignancy of Chandra's feelings:

> A lingering doubt besets the mind.... Does anyone hearken to my yearnings? Is my cry, a cry in the wilderness? Are the overtones of my voice heard on the other side? Such thoughts make one recoil and retire to solitude. Still emotions surge, eluding reason. An urge to kindle the flames of attachment persists. You want to reach out — you ponder, Is my absence felt, would my decision to stay back reflect disappointment?

It is with a sense of trepidation that I present these rather private exchanges between Chandra and his sister. I only do so to reveal the intensity and depth that characterized their exchange, mostly mediated through letters. On his last visit, Chandra took back with him the letters that he had written to my mother. "I want a part of Savitri to be remembered through my letters," he said. With again a gentle hesitation, he gave us a folder of the letters my mother had written to him during her terminal illness. "I thought this is a facet of your mother you would like to know" — and on the folder, in his inimitable handwriting, were the words "from Savitri during her last illness."

The writing we knew only too well from a different time. As my sisters and I were growing up in Madras, Chandra was Ayya Mama, the uncle who was far away. We cherished the books he sent us — Pinocchio, Grimm's and Anderson's Fairy Tales, Alice in Wonderland. To my sister Sudha came the books on science and mathematics to pique her curiosity. A book of Michelangelo's paintings for my mother. The beautiful penmanship of the inscriptions inside in the blue black ink was always a point of wonder to us then, as it continues to be now.

Stories abounded about the hero who was larger than life. His childhood stories, his seriousness of purpose and academic intent were mythical. The stories about (to quote my mother again) the "implacable patrician, out of bounds of trivia" reverberated continually in our lives and in the lives of our cousins.

But over the years, Chandra stepped out of the stories and became a real presence in our lives, particularly in the last few years. The visit in December 1994 left a deep impression on my sisters and me as he assumed a parental role in our lives. He wept with us, he partook of our pain. He spoke to us late into the night about isolation, of having lived between worlds, of Eliot and Chekov, of mythical Indian heroes, his mother and her literary pursuits.

The anecdotes still resonate. So too the intense emotions and respect we hold for our uncle Chandra.

Chandra in Focus

Sanjay Kumar

My first meeting with Chandra was way back in December 1986 when he came to deliver the Vainu Bappu Memorial Lecture at the Indian National Science Academy in Delhi. I phoned him later in the evening at his hotel and asked him if he could spare some time for an interview. He immediately asked me what specifically I wanted to interview him about — a question many people normally do not ask journalists. Fortunately I was prepared for that.

A journalist friend of mine working for a national news agency had approached Chandra for an interview after the lecture and Chandra had shot the same question. Unprepared, he made the mistake of saying that he wanted to ask him about the Steady State and Big Bang theories concerning the origin of the universe. Chandra told him that he should interview somebody else since he did not work in that area. Hurt at this dismissal, the journalist was livid at Chandra. But his narration of the encounter forewarned me. And I had a list of subjects ready, which appealed to Chandra, and he agreed for the interview and I was able to do a sufficiently long interview the next day. ("Why I Am Not a Believer," *The Hindustan Times*, 31 May 1987, and *Science Today*, January 1988.)

Sometime in 1989 I came to know about Chandra's forthcoming visit to India and decided to do some video recordings with him. I was seriously concerned that no good quality footage on Chandra existed, at least in India. Hardly any attention seemed to have been paid to this aspect — perhaps because Chandra was in the far-off US and was of no value for vote seekers here. I wrote to Chandra saying that I wanted to record interviews with him and also cover his programme in different cities of India. He did not reply. But we were given the impression that we could proceed.

Somehow there was a strong prevalent view that Chandra was very hostile to camera crews. "He throws camera crews out and will throw you guys out," was a refrain I heard repeatedly. I do not know the source of this rumour but this made us enormously insecure. The insecurity continued all through our tour to Roorki, Delhi, Bangalore and Pune, places Chandra visited and we covered. After all, it was a question also of putting a lot of self-raised, borrowed or donated resources at stake. And, at the end, what if he just threw us out...! We were apprehensive that we might annoy him without intending to.

Chandra did not display any hostility, but we were very cautious. We avoided putting any direct lights on him. It almost became non-interference in each other's affairs. As for the rumour, it is possible some incident had taken place in the past in which some insensitive camera crew might have gatecrashed during his lecture and without his permission had bombarded him with annoying heavy lights. I asked him if our shooting was causing him any trouble and he said there was no problem since the lights bounced off the ceilings.

Chandra maintained a formal distance with us, especially in front of others, and we reciprocated. He seemed to ignore our existence. He was certainly not camera-conscious. But he seemed to have a good sense of the impact his statements would have on the viewer. His long interview, interestingly, brought forth diverse shades of his inner feelings. When we met him otherwise, and we had several occasions — since we were following him everywhere — he was warm, friendly, jovial and polite.

I felt he could be informal, but only after some time, not in the very first meeting. His seeming coldness had an intimidating effect. At the same time, it seemed to create an aura of special mystique around him. I wondered, could he be outrageous or funny in public? Could he, for instance, have stuck his tongue out like Einstein and get photographed? I am sure many who have met Chandra would give an emphatic "No!" as an answer. But he was certainly not devoid of a sense of humour.

We had taken the same flight from Bangalore that he and Lalitha were on going to Bombay *en route* to Pune. At the Bombay airport, while the crew members and Chandra had gone to collect their baggage, Lalitha and I sat together chatting. When the luggage arrived, we proceeded towards them and on seeing us together, Chandra teasingly quipped: "Oh, so the bird is with the boy!" Both Lalitha and I burst into laughter while she said: "Just listen to what he is saying!" We collected our baggage and were about to part. Just then, Chandra came to me and shook hands, thanked me and said: "I am very grateful to you for keeping company with my wife." I was almost speechless.

Here I was, just a young boy and there was this old man and giant of a Nobel laureate expressing his gratefulness to me!

We, who are in the business of filmmaking, often encounter altered responses of individuals in front of the camera and off camera as also their apprehensions. How did Chandra react? I found some interesting instances that I wish to narrate.

I distinctly remember that just before we were to record his interview in the garden at his brother's residence in Bangalore, he called me aside and asked me what questions I was going to ask. I briefly described the issues I had in mind and told him that I had made a list. To my consternation, he straightaway demanded that piece of paper for reading. I was reluctant since I had only a rough scribbled set of questions (I was also conscious of Chandra's elegant handwriting *vis-à-vis* mine!). Besides I did not consider it a good practice to reveal the questions to the interviewee in advance. But he professorially demanded it, saying that it did not matter in what form it was. I had to force myself to give in.

He went inside the house and a short while later came out and called me aside again and waved the paper in front of me. He pointed towards certain questions and said in a somewhat exasperated tone: "You really want me to answer *these* questions?" "Yes," I meekly replied. He repeated the question in a more skeptical tone while pointing towards the paper, this time his words coming almost like a staccato fire: "You want to ask me what my reaction was when I got the Nobel prize. You know I was irritated! *You want me to say that!!*" He asked me to delete that question.

And he continued: "And here, you have..." There was some question about his views on the "reservation" problem in India especially *vis-à-vis* the scientific institutions. Chandra was not sure whether he should open his mouth on these controversial issues. (This was a period when crass populist measures were beginning to be unleashed by the then Indian Prime Minister V.P. Singh to appease the backward castes through reservation quotas — basically a vote-catching gimmick — for getting admission to academic institutions or for getting jobs — bypassing merit as a selection or promotion criterion. This set the whole country afire and finally led to Singh's downfall.)

I tried to convince Chandra that he should air his views for sure because the scientists here, being largely dependent on government funding, may not do so and his frank opinion would have its own value. And in any case we could decide at the time of editing whether to retain his "controversial" views or "censor" them or use them in some other form! He agreed and talked about the pitfalls of populism.

There were other controversial inputs. These pertained to his iconoclastic views on religion. He seemed pleased with the published version of his interview "Why I Am Not a Believer," a copy of which the Indian embassy had sent him. He told me in an amused tone: "Some people back in the US say — Chandra has atheistic–communistic views!" To me he seemed happy with his atheistic disposition. What he said in front of the camera about the Hindu holy text *Bhagvadgita* did not surprise me. But whenever I play back the recordings, I can't help remembering Salman Rushdie, in these times of the crescendo of Hindu fanaticism.

When I asked him if he had studied the *Bhagvadgita*, he said a few years ago he had spent several weeks reading it along with an abridged version of the *Mahabharata*. "I read it not with any prejudiced notion but in the manner in which I approach great literature. There are many things in it which I consider great literature, but somehow I was not at all impressed with Krishna. In fact, I found the whole attitude expressed in the *Bhagvadgita* as simplistic." Chandra said: "Arjuna raises again and again the question 'Why should I do this? Why should I kill my brothers and elders?' And Krishna says: 'You haven't killed them. I have killed them already.' 'What is my resurrection?' 'I'm your resurrection.' And Arjuna finally asks: 'How can I be sure?' And then Krishna shows his great form. I found that scene obscene. I am sorry to use that strong word. It was a way of essentially coercing him to believe. So, I am afraid I was not moved by the *Bhagvadgita*." "In other words," he continued, "I can read *Bhagvadgita*, recognize it as great literature, recognize that parts of it are wholly noble as instructions to people. But I am unable to accept it as divine and as one commonly says — a testament directly of God — because I know, at least I think I know, the *Bhagvadgita* was written by man and to attribute it to anything more seems to be irrational."

During the course of the interview on camera, I decided to exceed my brief and asked him what his reaction was when he got the Nobel prize. He did reply, but this time differently. He started: "To be frank I was disappointed." "Why?" I asked. "Well, what does it add? In my particular case it was awarded for some work which I did when I was a young man. I can remember what that young man was but it is not me."

He went on: "There is a wonderful play by T.S. Eliot called *The Family Reunion*, in which the lord of a manor returns after a long time to the manor.

The whole place was arranged exactly the way it had to be. And the people said you ought to feel comfortable because the place is the same. Then he (the lord) goes and stands by the window and sees Harry of old times but that is not he. He goes to a garden, smells the flowers, remembers how in earlier years he had stood in the same garden but the person there was not he. He felt very sad. And, in fact in the end, he leaves the manor."

Chandra's face looked grim. "So, I really do not know to what extent one can be sober — I mean sober in a very technical sense." He said: "One can think back about one's life and one sees that what one did at that time has received recognition. But what has it to do with me now? And suppose I had not received that recognition. Would posterity think less of me or more of me?"

Recollections

Anne Magnon

Midi l-haut, Midi sans mouvement
En soi se pense et convient soi-mme...
Tte complte et parfait diadme,
Je suis en toi le secret changement.

Paul Valry
"*Le cimetire marin*"

Midday up there, Midday motionless
Self-thinking, and self-dialoguing...
Complete mind and perfect diadem,
I am in you the secret motion.

Paul Valry
"*The Graveyard by the Sea*"

Introduction

Hazard, or necessity? I came into this world towards the end of the Second World War. France, my country, was but ruins and mourning. I still remember my mother describing my first months in the underground tunnel, a shelter built in the playground of our school to protect us from bombs. And I can vividly reconstruct her descriptions of the flashing rockets drilling the night skies above the nearby railway station, a few minutes before the bombings. Also, our house being occupied, the (naked) footprints left by enemy soldiers on the very high ceilings above our staircase were, and still remain, a puzzle.

With this most striking introduction to the mysteries of life and death, and of gravitation, I started growing, dressed for the years which followed, in the indestructible silk of the allied parachutes bringing food and medicines to men, including my father, hiding from Hitler's troops. When my legs became sufficiently strong, my father, who had recovered from his underground years, started taking me and my brother for endless walks, miles and miles in the countryside where devastated villages were still showing the wounds of the war. Here and there we would stop in some meadow where he would unfold his newspaper, or open his books to prepare his class; he was teaching philosophy. And we would play at all sorts of games; the favorite one was to try and repeat, without any mistake, various idioms such as "theskyisblue."

One day, as we were sitting in one of our favorite meadows, my father, still holding his newspaper, suddenly and in the most solemn fashion, started shouting: "S. Chandrasekhar, the famous Indian astrophysicist, has discovered a limit in the universe: the Chandrasekhar limit!!"

A limit in the Universe? That sounded like an origin, a creation! Had Chandrasekhar met God? We, the children who, a few minutes earlier, were still painfully struggling with "theskyisblue," became pale. A maelstrom started whirling around; that was the first and clear encounter with metaphysics and the anguish of Pascal. Our walks in the countryside quickly turned into astronomy initiation and my young imagination started carving portraits and figures of the famous man.

From the country of the forty thieves, and their treasures hidden in a magic cave, India soon became the roof of the world and the abode of gods. I quickly made up my mind: Chandrasekhar had to be some kind of Siddhartha, a prince able to practice contemplation and meditation, and, knowing past, present and future, to pronounce the sacred *Om*, to meet the *Atman*, the principle of life where soul, body and ego (mind and matter) can mingle. A wise and holy man, at the source of Reality.

Deep in my subconscious, I must have started craving for the mysterious initiation, the walk towards the Man of the River able to teach you that streams have no beginning and no end, that time does not exist, that everything is in the present.

The Encounter

Thirty years later (hazard or necessity?), I landed in Chicago (June '74), as a young associate professor and visiting scholar in the Relativity Group that had just gathered around Chandrasekhar. Of course nobody there knew the above details, but it is to be expected that, unlike me, many people at the Enrico Fermi Institute had heard of the Cambridge years, of the many stories surrounding the famous controversy between Chandra and Eddington about the equally famous Chandrasekhar limit and the fate of stars. (I read about all that much later, in the superb biography written by K.C. Wali, who must be thanked, here, for inviting me to contribute to the present volume.) From Cambridge to Chicago, Chandra had kept devoting his life to astrophysics and the study of stellar evolution. When I arrived, he seemed to be mostly occupied with the black hole stage of the gravitational collapse of stars.

A few days after my arrival and still remembering the holy man of my childhood dreams (sitting in meditation in his cavern, wearing a saffron gown, among garlands of flowers and glittering candles), it was, needless to say, a real shock to meet him at the door of the Institute, getting out of his enormous air-conditioned Buick, in a perfect gray silk suit. He was carrying the preprint of a paper that he had to referee for some journal. He gave it to me, asking me to report on it before the end of the week.

Most anxious, I took the paper to my room on Kimbark Avenue, and started working on it. It was a long paper on the Schwarzschild solution. The author was unraveling the asymptotic properties, symmetries included, in connection with the black hole horizon. After some struggle with his calculations, I gathered that he was aiming at some uniqueness theorem. But strangely, it made me somewhat happy that he had no comment on the Schwarzschild singularity. I had, and still have, grave reservations about black hole singularities. I consider them as pathologies that signal limitations of our geometrical approach to the universe.

When I was ready with the paper, I went to Chandra and presented a summary. Chandra's reaction was extremely brief. Saying, "I see: just the Birkhoff theorem," he disappeared. Chandra did not look very satisfied either with the paper, or with my comments. Maybe, the subject from white dwarfs to neutron stars and the ultimate fate of collapsing matter was, though fascinating, a painful one for him. I was about to learn that he was, in fact, much more involved with the Kerr solutions, and exact models which, according to his comments, were "made in Heaven." My feelings towards exact and geometrical models are unfortunately the same as those I have about singularities: there must be something beyond, there must be deeper principles.

Are we a facet of the mental activity of the Creator? Can a creative, and omnipresent, principle account for the interconnectedness of that part of Reality which seems to be accessible to us, through perceivable tracks only? Are "we" setting the Foucault pendulum in motion, through our ability to filter only a finite volume of information? It appears to me that at the source of such speculations could be the *Chandrasekhar limit*: an intuition that the ultimate fate of matter would also raise the question of the intelligibility of the world, and the related phenomenon of perception, shielding us from various abysses. With all due respect to Einstein, there must be subtle limits in the Universe which make it *comprehensible.* The mysterious force or principle holding everything could be a facet of the creative "Supreme Spirit" beyond geometrical methods.

Having met Chandra, the legendary and monumental figure in science who taught so many things about stars, the vision of my childhood should have been shattered. But it is not so. India still is the roof of the world. The vision is persisting, as a crystal-clear proof that life can build on archetypes. For me, Chandra is in fact more than that. Maybe a myth, a symbol, as intriguing as the problem of Reality and the phenomenon of life and death. As far as I can search in my recollections, the smell of earth after a warm rain has always been irresistible to me, just like hibiscus flowers in an Indian garden.

What Knowing Chandra Means to Me

Takeshi Oka

When I started to work at The University of Chicago in 1981, one of the first things I did was to visit Chandra in his office. Ever since Gerhard Herzberg introduced me to Chandra's writings, I had been an avid reader of his articles on science and scientists. The door of his office was wide open, and he greeted me most graciously. Since then I had been greatly inspired by many conversations with him. As well known, he had a phenomenal memory and an inexhaustible supply of anecdotes about scientists from his own association and from the history of science. We met for lunch perhaps once a month, and the number of our conversations was easily over 100. He never engaged in idle talk, and he never repeated the same story. There were his favorite subjects: general relativity, Newton's *Principia*, the Indian mathematical genius Ramanujan, Russian scientists, and others of which he talked about many times. In such cases, he remembered where he had left off the previous time, and he would resume from there, even though sometimes there were a few years intervening. Each time he opened a new door and introduced me to a new vista and surprise.

He had a unique perspective on the assessment of scientific work. He once stated, quoting his old friend, Milne, that "posterity, in time, will give us all our true measure and assign to each of us our due and humble place; and in the end it is the judgment of posterity that really matters." He then went on to say, "And it is well to remember that there is, in general, *no correlation* between the judgment of posterity and the judgment of contemporaries." There are countless examples of this last statement in art and music and in the history of mathematics and science, but one wonders whether this still applies for contemporary science. Chandra was reticent in criticizing scientists but judging from his respect of some contemporary physicists, I sensed that his judgment was quite independent of the current consensus. The most precious thing that I learned from these conversations was the sense of the eternity of human endeavor; in Wali's words, "magical feeling of oneness with the past and the continuity of human effort."[1] Here was the man who learned from Kepler and Newton, continued the work of Jacobi and Riemann, Rayleigh and Poincaré, and tried to reach for posterity. I am shortsighted and often go astray in the ups and downs of science. Sometimes I get depressed when I "find myself forced to listen to pompous and tiresome people," but after such experiences, conversations with Chandra pumped me full of oxygen and saved me from suffocation.

Most of the time he was talking and I was listening, but he also was a good listener. His grasp of various domains of physics was so extensive that he could perhaps adjust to the interest of any physicist. Many times he started our conversation by asking me what's new in my laboratory or observatory. It was very encouraging because he not only followed the technical aspects of my work, but he also appreciated the meaning of it. He was also interested in my students. Once I told him that I had excellent students and some of them were better than me. He said, "When you were young, you admired those great scientists and thought they were out of your reach. Now you are older, and you admire your students and say they are better than you. You never win!" He said that with a twinkle in his eyes.

He was extremely generous and a kind listener. Several times I contradicted him on certain crucial issues of science. He maintained his points and tried to refute but never once did he brush me aside as much lesser people have often done. Once he called me late at night and apologized that he had treated me harshly. I was astonished because I never felt that way. He was a warm, kind, and caring friend.

Now Chandra is gone, and this campus is no longer the same campus for me. But, in a peculiar way, I began to feel his presence even stronger and everywhere I went. Three weeks after his death, I was at the Kapitza Institute in Moscow. I saw pictures of Kapitza and Rutherford, Fowler, Dirac, and Mott, Chadwick and Blackett, and, of course, Landau, Lifshitz and many former Soviet scientists about whom Chandra talked so much. Returning to Chicago, I started to re-read his collected papers, monographs and biography. He talks to us so eloquently through those books! I am extremely lucky and privileged to have known him and that I am able to continue learning from so many materials that he has left for us.

Note

1. *Chandra: A Biography of S. Chandrasekhar* (Univ. of Chicago Press, 1991), p. 26.

Chandrasekhar — Some Reminiscences

S. Ramaseshan

Introduction

Since Chandrasekhar first left India in 1930 I have met him many times. We have exchanged many letters over the years. Providence showered on me an unexpected gift — we became very good friends. He wrote once:

> I should not be writing this letter, if I were not sure of your long-standing friendship.

In this essay I relate a few incidents which took place; but memory can be quite deceptive. When I use the active voice, the words may not be the exact ones spoken at that time. I lace the stories with my comments — which the reader may ignore.

The Last Conversation and His Death

My last conversation with Chandrasekhar was in July 1995. I called him up from Canada on my way to a conference in Montreal. We talked for a long time about many things. He said, "I have finally received the printed copy of my book on Newton's Principia, I like the cover." He had worked on this book for almost four years. "My work is finished," he said. On hearing that I was not going to Chicago this time, he asked me in what I felt was a wailing voice, "Sivaraj (that is, me), aren't you coming to see me?" That question still haunts me — for within two months he was gone.

I was quite ill when in the early morning of August 22 I got a telephone call telling me that he was no more. I phoned a few friends in the USA. I could gather that he had a slight pain in the chest on the morning of the 19th. As it was on the right side he and Lalitha were not too worried. I am told that they even discussed what they should do on the occasion of their 60th wedding anniversary, which was to come a year later in September 1996.

On the morning of the 21st he drove all by himself to the University Clinic to get himself checked. Doctors found that he was having a massive heart attack. One conjectures that he must have had the attack during the drive. They phoned Lalitha. She was with him for some time. He was heard to tell her, "I think it is the time for you to start making the preparations we had planned." He died during the day. He had stipulated that there should be no memorial service or any meeting. He was to be cremated and no one was to be present; and the ashes should be brought to Lalitha. He had said to me, "My work is finished"; it seemed to me as though he had arranged with providence for his end. As Plato said of his master, the great Socrates:

> Such was the end of a man who I think was the wisest and the justest and the best man I have ever known.

Polarisation of the Sunlit Sky

In 1951 when Chandrasekhar visited Bangalore he came for lunch to my little house at the Indian Institute of Science. I had recently been married. He was full of laughter, quipped with Kausalya (my wife) and discussed with her sari fashions. He told us many humorous stories about himself and other scientists.

The next day he gave a lecture, "On the Polarisation of the Sunlit Sky." It was perhaps the best scientific lecture I have ever heard (probably because I had some background in this field). It was the pinnacle of his monumental work on radiative transfer. He told us how he got interested in the problem of the atmosphere of a star; the need to treat the coupling by photons of the different layers with varying properties; how the principle of invariance evolved following the work of the Armenian astrophysicist Ambartsumian; and how he developed "invariant embedding" with the intention of applying it to explain the polarisation of skylight — the radiative transfer in a finite slab of air which scatters the sunlight in the sky. He met with the problem of representing polarisation of light. He went back to the great masters and discovered the appropriate parameters from Stokes' paper published in the late 1870's. Using these he formulated integral equations for multiple scattering which were so complex that even the best mathematicians would have left them alone. To quote Rajaram Nityananda, "He talked to these equations personally and intimately till they gave up their secrets to him." At the end of the lecture C.V. Raman, who was in the audience, commented, "Rayleigh formulated the theory of light scattering to explain the blue of the sky in 1871. For eighty years the problem of the experimentally observed polarisation of skylight defied the efforts of the great masters. Today we heard a most extraordinary exposition. How the problem was solved by a combination of physical intuition and powerful mathematical analysis." Then he asked, "Professor Chandrasekhar, can you please tell us how long it took you to work out the results you have published in the four-page paper in *Nature*?" Chandrasekhar, after some thought, replied, "More than two-and-a-half years." That issue of *Nature* (1951) had not reached Bangalore then. But Chandrasekhar, as was his custom, had sent a preprint to Raman by airmail.

I jump 35 years — as this relates to another Indian who was also an expert in the same field. Chandrasekhar wrote to me:

> Stimulated by the handsome references to Pancharatnam by Michael Berry, my colleagues and I are interested in the *Collected Works of Pancharatnam*. I should like to buy a copy.

I sent him a complimentary copy. He being the High Priest of Polarisation Optics of this century, I asked him whether he would be inclined to write for the Pancharatnam issue of *Current Science*. He wrote:

> It is almost 45 years since I thought of the Polarisation of Light. I may not have anything worthwhile to say. However, I always congratulate myself that when I read Pancharatnam's classic paper long ago, I did recognise its merits.

Space does not permit me to write about my meeting him in Yerkes in 1954; his accounts of his collaboration with Enrico Fermi; Fermi's death, which had occurred just a few weeks earlier (he told me the now well-known story of the elephant); how he decided and took the US citizenship in 1953 (without consulting his father). "Now Lalitha has been appointed the peacemaker," he said. "You are a coward," I said. He related incidents of his and C.V. Raman's encounter with colour prejudice in the USA and many other stories. I shall never forget this memorable meeting.

Wali's Biography and Raman

Chandrasekhar asked me once what I thought of Wali's biography. "It is a remarkable biography. It reads like a novel." "Yes, I read your views in *Current Science*. Do I detect a slight undertone of disapproval?" he asked. "Yes, I do not like those parts where by implication it is derogatory to you."

"Derogatory to ME" — the Me became a high monotone, reminding me of his uncle. "When Raman offered you a position at the Indian Institute of Science, you wrote him a very fine, courteous and what I consider to be an honest letter; that you were flattered, you were not sure whether you would fit into an experimental group, and finally you were afraid that the media attack against Raman would increase if you, his nephew, were appointed. And this might affect your performance as a scientist. Therefore you were with great regret not accepting his offer. Wali, on the other hand, has written it in such a way that one gets the impression that you were responding to your father's wishes to stay away from Raman. This implies that your letter was a false one — that you were dissimulating. I have now known you for a long time. I cannot imagine you dissimulating or playing the hypocrite."

Chandrasekhar was silent for some time and said with a smile, "I did not write the biography, it was Wali who did."

I told Chandrasekhar, "I am told you are a bit ambivalent about Raman. In your letter, when Lalitha and you presented Ramanujan's bust, you say:

> as a companion to the bust of Raman so that the bust of the greatest physicist of India could be along with that of the greatest mathematical genius of our times, who happened to be an Indian.

"Do you really think Raman was the greatest physicist of India?"

"Of course I have no doubt on that score." People in India remember him for the discovery of the Raman effect, which won him the Nobel Prize. One great discovery advances science, but it cannot be the criterion for judging the quality of a scientist. Raman's work on acoustics is first rate — his work on the violin is considered a classic even today. Acousticians wonder how he could have done this work in those days. In optics he was a master. It is a great pity his *Lectures in Physical Optics*, which I used for giving an advanced course in optics, is not so widely known. His Nobel Prize, in my view, retarded his science, but I may be wrong. Even so his later work on Brillouin scattering and crystal transformations was again world class. One feels sorry he committed major scientific errors in his later life — and he became aggressive too. Unfortunately this is what many remember him by — not the excellent physics he did earlier. He was a brilliant classical physicist who made a spectacular discovery of a quantum-mechanical phenomenon which influenced physics, chemistry and even technology, as he predicted in his Nobel Lecture.

"How do you rate Raman by world standards?"

"As an experimental physicist?"

"Yes."

"For that one has to compare him with men like Rutherford and Fermi — and the answer is obvious."

"If that is the yardstick you use, would not all experimenters except perhaps one or two, like Michael Faraday, fall by the wayside?" After some thought he remarked, "Yes, that may probably be true."

Raman and Chandrasekhar met only a few times after he left India. From all accounts the meetings were very cordial, often hilarious, each poking fun at the other. Raman proposed Chandrasekhar for the Fellowship of the Royal Society in 1942 and he was duly elected in 1944. Raman insisted (in 1948 and 1949) on putting him up for the Nobel Prize for his early work (1930–35). Chandrasekhar consistently refused him permission to do so.

After 1961 one did perceive a certain coldness on the part of Chandrasekhar towards Raman. The reason for this seemed clear to me when Chandrasekhar told me the following story. I reproduce the story as given in Wali's book (Wali, page 253).

I remember one conversation with him (Raman) which throws a characteristic light on him. In the last week of November 1961, when I arrived at Raman's office, he was unwrapping my "Hydrodynamic and Hydromagnetic Stability", which had coincidentally arrived by mail just then.... He turned to me and said, "The only book of this size I have seen before is a novel by Anthony Trollope — absolute trash."

When he related this to me (much before Wali wrote his biography) I said, "The words 'absolute trash' obviously were a comment on the novel and not on your book." "Why do you say that?" he asked. I remarked, "I grant that Raman could be very insulting, but it is impossible for me to imagine him insulting you (or for that matter Harish Chandra or Pancharatnam). Chandrasekhar said rather coldly, "You have the privilege to hold on to your views." Chandrasekhar could be quite unforgiving towards anyone who derided or made fun of his scientific work.

In the third week of August 1995 (thanks to the efforts of Krishnamaraju of the Raman Research Institute (RRI)) a bunch of letters exchanged between Raman and Chandrasekhar during 1933–61 was found (S. Ramaseshan — "S. Chandrasekhar and C.V. Raman — Some Letters", *Current Science*, 1995, 70, 104). The last one I reproduce:

9th December 1961

My dear Chandrasekhar

A few days ago I received from Oxford University Press a presentation copy of your treatise on Hydrodynamics and Hydromagnetic Stability... . The magnificent way in which it has been printed, illustrated and got up is beyond all praise. The beautiful photographs and drawings interspersed throughout the volume would attract many readers who might otherwise have been put off by the rigors of the mathematical analysis. You have rescued from oblivion the work of many investigators in this field which might otherwise have remained buried in the learned periodicals. They have reason to be grateful to you for surveying their results and presenting them along with your own thoughts and contributions. It is remarkable how many different fields of study the book illuminates and is likely to illuminate during the years to come. I am grateful to you for the presentation.

Yours affectionately
C.V. Raman

This was directed to Chandrasekhar's Madras address. Although he was in India at that time I found he had left Madras by then. It is my conjecture that this letter either was not forwarded to his US address or was missed in post. For Chandrasekhar, who was so punctilious in replying to letters, had not replied.

He Enters General Relativity

A friend of mine was visiting Chandrasekhar. They were walking down the corridor. Suddenly he saw a young man wearing torn jeans coming from the end of the corridor. Chandrasekhar was of course immaculately dressed — in his charcoal-grey suit — handsome as ever. He straightened his tie, went forward at a fast pace, greeted the young man, shook his hands, talked to him for some time, opened the door of the car almost with deference, and the young man got in and drove off. Chandrasekhar came back and joined my friend, who asked, "Who is that hippie?" Chandrasekhar said, "Don't you know, he is Geroch." "Who is Geroch?" "He is one amongst the most outstanding relativists of today — he is in his twenties. I have invited him to lecture in Chicago this summer. I am attending his course. It takes me the entire weekend to work out and understand some of the equations he puts on the blackboard — to be ready for his Monday lecture. When I am stuck I ring him up, even if it is late at night, to bail me out."

At 55 he was humble enough to learn relativity from youthful scientists; at 65 he held his head high amongst the brilliant young relativists. At 75 he was respected by both the young and the old for his unparallelled contributions to general relativity.

It was about this time that Chandrasekhar told me,"The trouble with Einstein was that he did not believe sufficiently in the general theory of relativity."

I honestly felt that he was going gaga. Many years later, after he wrote his monumental treatise on black holes, I told him of the fears I had. He laughed loudly in his characteristic way.

It was later that I understood what Chandrasekhar meant when Abhay Ashtekar explained to me:

> If Einstein had believed in his general relativity sufficiently and deeply enough, he would have worked out the consequences of the non-linear aspects of the theory in the strong field regime. Chandrasekhar believed that Einstein was the only person who could have done this at that time. Then general relativity would have entered the mainstream of physics four decades earlier.

Chandrasekhar told me, "Many authors had not considered the gravitational radiation reaction in general relativity. When all my old problems were reexamined very unexpected results turned up. For example, I was able to show that all rotating stars are unstable due to radiation reaction."

Abhay Ashtekar says, "His intuitive feeling that the radiation reaction in general relativity is important, is one of the deepest insights in relativistic astrophysics."

Music, Literature and the Arts

Chandrasekhar's appreciation of music, literature and the arts was staggering. He mastered each of these fields in his characteristic way. He heard each of the symphonies of Beethoven and the operas of Wagner many times. His attachment to Western music was both intellectual and emotional.

Often he read out to us pieces from poetry and prose which he liked. His readings from T.S. Eliot or Hermann Hesse were delightful. When he read passages from Virginia Woolf in his modulated voice you felt you understood this rather difficult author. But I think it was Shakespeare that he loved most. He must have read every play of Shakespeare at least twice. He could lecture on any play better than any Oxford don. I felt that Shakespeare gave a sort of direction to his life. He often told me that reading *The Tempest* helped him in the attitude he had in doing research. I can easily imagine Chandrasekhar to be Prospero of *The Tempest.*

Like Prospero, the grievance he nursed vanished when he did his science (Shakespeare calls it the enchanted arts) and he became free. However, I think what finally liberated him and his soul was his complete surrender to his god — whom he called "the unmatched and unmatchable Newton".

It would be impossible for me to convey to you the excitement of the exposition he gave us on Newton's Principia on one of our visits to Chicago. Almost true to his chosen title — "Newton's Principia for the Common Reader" — he lectured to Kausalya as though she was the Common Reader. I felt that he was really constructing the sentences that he was to write in the introduction two years later. We cherish a photograph of Kausalya and Chandrasekhar which I took. He is holding Cajoris' 2nd Edition of the Principia, which he had brought for eight thousand dollars.

Once when we were in his drawing room Kausalya asked, "Is that a Monet?" "Yes — only a print — I cannot afford an original." He told us that a chance remark by Roger Penrose that his excursions into general relativity reminded him of the series paintings of Monet, triggered him off to study everything written about Monet and by Monet. He collected excellent prints of all his paintings — especially the Series Paintings of the Grain Stacks, Poplars and Mornings on the river Seine. His word pictures of these paintings were so graphic that he made us see things we would never have seen:

> I feel I am exploring the landscape of the mathematics of general relativity as Monet explored his Haystacks which took new forms in the changing colour of light during the day and during the different seasons.

We sat there with our mouths open. It was an entrancing half hour. We can remember almost every word he spoke. Did we understand whatever he said? We do not know. His enthusiasm, his sparkling eyes, the cadence of his voice remained with us for a long, long time.

Basilis Xanthopoulos

My visits to the USA were never for more than ten days. A day in Chicago was a must. I phoned him once. He said, "I am flying over to meet Roger Penrose this evening. Can you come after three days?" I went on the specified date.

He was at that time probably working on black holes and cosmic strings.

"Whenever I meet a stumbling block I go and meet Roger Penrose. We spend an hour together in the morning when I present him with my problem. We have four to five hours of discussion after lunch. Then at dinner we talk of other things — and I fly back. In no case has he not cleared up my doubts in physics or in mathematics. An amazing man."

During the period when he collaborated with Basalis Xanthopoulos in Crete and Valeria Ferrari in Rome, he met them often — in Chicago, Crete or Rome. It was with them that he discovered the underlying similarity between the mathematical theory of black holes and of colliding gravitational waves. We at the Raman Research Institute (RRI) had the privilege of listening to his latest work which had not yet appeared in print. On every occasion he mentioned with gratitude and affection Basilis Xanthopoulos and Valeria Ferrari.

So when I read the horrible story that while lecturing at his University, in Crete, Xanthopoulos was murdered, I wrote to Chandrasekhar a condolence letter; he was touched and wrote, "My association with Basilis was the most enduring personal relationship I have had in all my 60 years in science." He also sent the foreword Basilis had written for Volume 6 of his collected works, which I also printed in *Current Science.*

When we met the next time I referred to Xanthopoulos' death. I could see in his eyes the agony that I saw in my father's when he lost his beloved son. I realised then, more than ever, that persons like Basilis Xanthopoulos and Valeria Ferrari were truly his children — children that he and Lalitha never had.

Return to India?

The following episode extended over four or five meetings in Bangalore, Chicago and Cambridge. I will try to give a connected account.

One day I asked Chandrasekhar, "In an essay you wrote about 30 years ago you said: 'I have been a foreigner all these years except for a brief vacation for eight weeks when I was in India. By "foreigner" I mean that I have never been allowed to feel otherwise. Yet people from other countries migrate to distant lands, build homes on new soils, adopt and are adopted by their neighbours. But why is it not so for me, an Indian?' " I then asked, "Has that feeling changed now?" "Unfortunately, no. I thought it would when we became citizens of the USA in 1953. But even today I feel at home only in India. I think it is a great pity — why do you ask?"

"If I offend you please stop me. Have you ever thought what will happen to you if Lalitha dies — and what will happen to Lalitha if you die? For me the US can become a very, very lonely place."

"Have you a solution?"

"Yes, I have. Why don't you come back to India? Bangalore would probably be the best place. You can be at the Raman Research Institute. You and Lalitha can stay in a cottage provided by the RRI Trust. It can also provide you with help, transport, etc. You can invite any of your collaborators to Bangalore. If you so desire, you can give courses of lectures now and then. I can assure you that very bright young people — and they are plentiful in India — will swarm around you. You can select the best of them to work with you. You would also be fulfilling Radhakrishnan's dream of having a Gravitation and General Relativity Group."

Always the gentleman, he said, "It is most gracious of you to have thought of me and to have invited me. The offer is very attractive. I shall surely think about it."

I informed Radhakrishnan that there were chances — even though slim — of his accepting our offer. So a part of the new building was redesigned — special permission was obtained to install a lift, for Chandrasekhar had had his share of heart attacks and heart surgery.

He told me later that he had given some careful thought. He felt that Bangalore would be a very good choice — "The only other place I could have considered was Oxford, because Roger Penrose is there. But there are some difficult conditions. Firstly, I would not like to give up my US citizenship — for it was the US that gave me conditions which made my work possible — conditions which I could not have got anywhere else. Secondly, I am not a rich man. I would get a pension. We need money to travel to conferences, to meet colleagues and collaborators. I am told that the Indian tax laws are quite formidable."

I met persons at the highest levels. I was assured that Chandrasekhar could retain his US citizenship, even if he lived in India. I was also assured that steps could be taken to see that he was exempt from paying tax during his lifetime. Even an annual grant could be given to RRI to cover his research, travel expenses, etc. I conveyed all this to him. At that time he had finished his work on black holes and had started his work on the collision of gravitational waves.

In 1983 he was awarded the Nobel Prize. I wrote him a congratulatory letter, saying that my children were also ecstatic, and RRI and indeed the whole country was thrilled about it. A month later, in January 1984, he phoned me to say that he would be accepting my invitation to give the inaugural address at the Golden Jubilee of the Indian Academy of Sciences, and that he was preparing an essay on "The Pursuit of Science: Its Motivations", specially for the occasion (an essay which is now famous). I was thrilled. He also conveyed to me the sad news that after getting the Nobel Prize it would be ungracious and appear ungrateful if he should leave the United States. Again a perfect gentleman. He went on to thank me for all I had tried to do and apologized that all my efforts had come to naught.

Conclusion

I quote statements made about him:

"There were two faces of Chandrasekhar: the stern, aloof, difficult-to-approach person and the kindly, charming, laughing human being dispensing jokes and anecdotes."

"He had an inexhaustible supply of anecdotes — from his own association with scientists and from the history of science. He never repeated the same story to the same individual — for he had a phenomenal memory."

"Each time you talked to him he opened a new door for you and you saw a new vista — a new surprise."

I was always flattered that he listened so intensely to one such as me. When I look back I shudder to think that I lived when he lived, that I knew him and he befriended me. For me this was the only proof of the theory of karma. I must have done something good long, long ago.

About his science he said once, "I work on my own for my personal satisfaction generally outside the scientific mainstream." It always became the mainstream a few years later. His earliest work led to the concept of neutron stars and black holes. His survey of Brownian motion started new fields, many not at all related to astronomy. His concept of dynamical friction has become as much a part of the vocabulary of astrophysics as the Chandrasekhar limit. Radiative transfer started a brand new field in mathematics called invariant embedding. His ellipsoidal figures of equilibrium have become important to fast-rotating pulsars. One can go on and on.

His prescience in taking up any problem makes one gasp. Believers would say, "He was touched by the hand of God — a touch that remained with him till he died." Such a statement would have made him extremely uncomfortable. But there is a simpler explanation — for him doing science was like breathing. He did what he could not help doing. As the poet said:

> As in yonder valley the myrtle breathes its fragrance into space.

Subrahmanyan Chandrasekhar Remembered

Vatsala Vedantam

In one of the last interviews before his death last summer, the Nobel laureate spoke to the author at his University of Chicago office.

"How can you write about me — you don't know me," he says.

"We have met twice and you have been talking," I answer.

"That still does not mean anything — you don't know my thoughts," he persists.

I try another line.

"Maybe I could talk to people who have known you?"

"Listen," he says, "I am asking an academic question. If you meet a person for a few hours, and talk to some of his associates, can that give you insights enough to write about him?"

"I am not writing a biography," I counter. "I am just setting down my impressions in a newspaper..."

He interrupts with: "How will that interest your readers?"

"That is how I profiled M.S. Subbulakshmi," I tell him. And add, "My readers liked that piece — including you."

He is unfazed.

"Subbulakshmi was different," he states. "You had known her all your life."

"No, I didn't," I protest.

"You knew her since your childhood," he continues, unperturbed.

"You had seen her act in the movies; you had heard her sing in concerts. You grew up with her. Don't you see, you were attuned to your subject?"

I give up. I have no more defenses.

"But you don't know me." We are back to square one. "You are not familiar with my work. You have not read my publications. You have not heard my lectures."

I interrupt feebly.

"I did hear you once."

"Where and when?" comes the swift question.

"At the TIFR auditorium in Bombay," I tell him, "I think it was in 1971."

"Did you understand what I said?"

"Not a word," I confess, but go on to add:

"I only remember how you ended your talk with that story about the Pandava princes and Arjuna's powers of concentration while looking at the eye of a bird. You mentioned that is the kind of concentration which a scientist needs to have."

The old man suddenly relaxed and smiled his rare smile.

"Ah, that was my favorite anecdote which I used to tell my students."

He leans back in his chair. The sparring is forgotten.

* * *

That was Subrahmanyan Chandrasekhar. Scion of an orthodox Shaivite tradition. Distinguished Service Professor Emeritus. Indian scientist. American Nobel laureate. The man who unraveled the mystery of the stars. The astrophysicist of the 20th century.

Young scientists say they feel "intimidated" in his presence. His colleagues maintain a respectable distance. Deans of of the faculties mention his name with deference. At the University of Chicago — where he has outlasted six university presidents —"Chandra" has become an institution in himself.

* * *

Debonair at 83, a distinguished-looking man dressed elegantly in a charcoal gray suit walked down the corridor and stopped at the door of his secretary's office.

"I am sorry to have kept you waiting," he apologized.

The Nobel laureate was exactly four minutes behind time!

As we stepped across the corridor to his office, he casually peeled off a note-paper from the door, saying, "I don't think you saw this."

It contained a brief message: "I shall be back in a few minutes. Please wait inside. The door is open — SC."

That piece of paper spoke more eloquently about Chandrasekhar than anything I had read or heard.

It seemed an awesome room. Portraits of Ramanujan on one side, Newton on the other. Statuettes of Socrates, Aphrodite, and Aristotle standing alongside Michelangelo's Moses dominating the rest. Volumes of books everywhere. The man of science is also a meticulous individual. He carefully places his black umbrella on its rack, hangs his coat on another, and fusses with the chairs.

"Are you sure you are comfortable?" he worries before seating himself in his armchair.

"Now, tell me," he begins.

You feel totally inadequate in that milieu and stammer something about wanting to write a human interest story about a great physicist of our times. He senses your discomfort, leans forward, and observes very earnestly:

"I hope you don't write anything like this: 'In July 1930, a young Indian stood on the deck of a ship, bound for England and higher studies. He was none other than S. Chandrasekhar, the future Nobel Prize winner!'

"That is how someone wrote about me once," he added. He couldn't have thought of a better way to start an interview.

But Chandrasekhar refuses to be drawn into personal conversation. In his own words, he is a highly private person — a loner with few friends.

"It is not a criticism," he adds. "It is only the character of my work."

At the same time, there is a note of regret when he says, "Many young people in this university are not even aware that I exist."

University students, on the other hand, claim that they are so awed by his presence that they even avoid crossing his door for fear of running into him. Says a research scholar from Bangalore: "He is so distant, I feel afraid to approach him." And yet, according to his colleagues, Chandrasekhar loves young people — the undergraduates — and would have them come and talk to him.

* * *

How does one profile an enigmatic personality like Chandrasekhar's? This "glorious scientist-in-exile, whose sparkling intellect can take you out of yourself and transform you into another world," according to Kameshwar Wali, his uncritical admirer and biographer. A scholar who is described somewhat exotically by fellow scientists in India as "a *rishi* who has chosen the difficult path of *sadhana* and *tapas* to attain his goal." A teacher who can bow in deference to a 19-year-old undergraduate whom he considers "one of the great scientific minds of this century." A humanist who can relate warmly to a high-school student in Bangalore when she proudly shows him her science project. A concerned host, who insists on making the beds for his guests when they visit his apartment in Chicago. A boss, on the other hand, whose secretary quit in anger, because she found him "arrogant, secretive, and frigid."

Mavis Lozano's main grouse was that the distinguished professor for whom she worked between 1975 and 1981 did not encourage friendship or familiarity. Even though she secretly admired this "fantastic mathematician" who returned to work within a month of an accident and a hip fracture, her American upbringing could not comprehend his aloof manner and formal behavior. However, she grudgingly admits that he never argued, never lost his temper, nor showed the least discourtesy to those around him.

"I disliked him all the same," says Lozano emphatically, in a separate interview. Reason? "He was so formal. Would you believe it, he wore a white shirt, a dark suit, and a tie all the year round? He did not remove his jacket even once in the office. Imagine being dressed like that on an American campus, where our professors work in shorts and T-shirts!"

To people like Lozano, Chandrasekhar would have seemed an anachronism — a total misfit in Chicago's racy ambiance. People like Lozano would also not understand why her "arrogant and secretive" professor should go to great lengths to obtain her address for a visiting journalist.

It is not easy to draw a portrait of a man whose personal life has been completely subdued by his brilliant scientific achievements. Educator, researcher, scholar, author. The recipient of the world's highest awards and honors. Perhaps the greatest astrophysicist of this century. Understandably, his acquaintances become adulatory when they talk about him. One of his close associates went as far as to say, "Chandrasekhar cannot tolerate trivial conversation nor does he suffer presumptuous fools. He, therefore, enjoys very few friendly relationships with people."

And yet, when the distinguished scientist visited the newly established planetarium in Bangalore a few years ago, he not only sat through an entire program, but later submitted himself to the absurd antics of an amateur photographer, who tried to make the great man smile for a picture!

Chandrasekhar himself feels that no one — except perhaps Kameshwar Wali — has really understood him and his work. Yet, to a nonscience person, he comes out as the gentlest of human beings. A truly civilized person, totally devoid of conceit or condescension. A man to whom courtesy and good manners are not mere appurtenances. They are all part of his great humanity.

The "unapproachable" man of science is in fact endowed with great humility. In 1968, when he was awarded the Padma Bhushan along with M.S. Subbulakshmi in New Delhi, Chandrasekhar crossed the stage, bowed in deference to the great musician and told her, "Madam, I am deeply honored to have shared the same platform as you today."

The best assessment of Chandra comes from himself. Seated on the lawn of his brother's house in Bangalore recently, he summed up his earlier remarks about himself and his work.

"I do not consider myself as someone who is striving toward a goal."

He measured every word that he spoke:

"In my perception, science is not finding solutions to problems. Nor is it making a discovery. To me, science is a search for truth, a search for a definite pattern in creation."

He was lost in thought for some time. Then went on softly, talking as if to himself:

"Mozart looked for it in music. Monet saw it through a haystack that he painted. Michelangelo found it in stone. Poets down the ages have written about it. Perhaps a scientist seeks it through a mathematical equation. But we can all see only a part of this pattern at a time. It is given only to a few to see the whole..."

It now becomes clear why this man of science has gone through the painstaking process of studying and understanding art, music, and literature. As one of his friends pointed out: "Chandra seems to admire all those great achievers who disciplined their minds (like himself) and chose the difficult path to self-realization."

When Chandrasekhar was invited to attend a Mozart concert in Europe, he made a detailed study of the life and work of the great composer. When his doctors advised him to take a break from research after a major heart attack and triple bypass surgery, he made use of the time to study Shakespeare's plays!

His 83 years are forgotten when he is absorbed in his work. Even though his vision in one eye is very diminished and wife Lalitha feels concerned about it, "I can still read with the other," says an irrepressible Chandrasekhar.

Maybe his sense of humor, his capacity to laugh at himself, have something to do with his amazing youthfulness.

To the last and inevitable question about the Nobel Prize, for example, Chandrasekhar replies with a mischievous smile:

"Have you heard the story about this general, who had all those gold medals on his chest? Someone asked him what they were for and he answered, 'This first one here was a mistake, the rest simply followed.' "

* * *

"I had no mentor"

"You came to America in 1936. Do you think you would have achieved what you did had you stayed back in India?"

"In a narrow sense, the answer is NO. There were better facilities for work here. I was also disconcerted with science politics in India. I was very sensitive and I desired the mental peace to do science the way I wanted.

Secondly, how can one evaluate scientific achievement? It is not a personal accomplishment. I had many students and collaborators. Science has to be an integrated effort. Otherwise, it would be too narrow."

"Who was your earlier mentor? And who influenced you most in your career?"

"I had no mentor. And nobody 'influenced' me. I wrote my thesis on my own. I have always been alone. This is not criticism. It is the character of my work."

"Was your father a dominating influence in your life?"

"All Indian fathers are dominating!" [With a laugh]

"Do you recall your mother and her attitudes, which may have shaped yours?"

"Yes, I recall a particular incident, which revealed my mother's extraordinary awareness. I was hardly ten years old, when she woke me up one morning and said, 'Do you know Ramanujan is dead? It has come in the newspaper.'

The very fact that she realized that Ramanujan's death was an important event showed her enlightenment in these matters. Her attitudes did influence me a great deal."

"Has your wife been a great support to you in your scientific career?"

"I have mentioned Lalitha in my book *Truth and Beauty.* My biographer Kameshwar Wali has also written a whole chapter on my wife. [Suddenly with a smile] Do you know the American press called that the best chapter?"

"Have you, at any point of time, regretted your decision to leave the country of your birth?"

"There is no point in regretting or being happy over decisions you have made. I think it's irrational to regret the past anyway. You must reconcile yourself to the life you have chosen and lived."

"Would you call yourself a religious person?"

"No, I am an atheist."

"Do you enjoy teaching?"

"I always integrated teaching with research. They support each other."

"What is it that makes Indians achieve more in this country (America) than in India? After all, it is the same brain. Do you think it could be the academic climate?"

"I wouldn't judge achievement by awards. The quality of science in India is good, too. But I remember, in the 1930s, the great scientists of that country were in the universities. But today, it is not so. And, that is a loss."

"Whom do you consider as the great scientists in India?"

"I think [Homi] Bhabha has contributed more than anyone else to science in India. Another was [Shanti Swarup] Bhatnagar, who is sadly underestimated by those who work in his own institutions."

"Has your personal life been complete and happy?"

"That you should ask Lalitha — maybe I could have given more. [Pause] I don't believe that a scientist — a true scientist — can ever have a complete personal life. [Pause again] I sometimes wonder whether all that I did and accomplished in my lifetime — was it really worth it?"

Kameshwar C. Wali later interpreted this comment as: "When Chandra asks — Was it worth it? — he is not being negative. It is just an awareness, another dimension of realization which dawns as one gets older."

Chandra with Ramanujan's widow, Janaki Ammal.

Andrzej Trautman and Chandra.
(Courtesy Morek Holzman.)

Chandra receiving the National Medal of Science from President Lyndon B. Johnson, 1967.

Lalitha.

Prime Minister Indira Gandhi and Chandra, 1968.

Chandra and Lalitha on the occasion of Chandra receiving the Donnie Heineman Prize, 1974.

Chandra receiving the Honorary Degree at Harvard in 1979.

Chandra on the occasion of receiving an Honorary Degree at Syracuse University, 1987.
(From left: Chancellor Melvin A. Eggers, Chandra, Kameshwar C. Wali.)

Bimla Buti and Chandra, 1982.

Chandra and Lalitha in Bangalore, 1994.

Special Convocation, University of Roorkee, 1989.

Thursday Afternoons

George Anastaplo

> There is no becoming of what did not already exist, there is no unbecoming of what does exist: those who see the principles see the boundary between the two [being and non-being]. But know that that on which all this world is strung is imperishable: no one can bring about the destruction of this indestructible. What ends of this unending embodied, indestructible, and immeasurable being is just its bodies ...
>
> — *Krishna*[1]

I first came to know Subrahmanyan Chandrasekhar a quarter of a century ago through his gifted wife, Lalitha, who had been a student in adult liberal education seminars I have conducted at the University of Chicago for some forty years. He and I would talk from time to time, usually for a few minutes during the tea hour before the Thursday Physics Colloquium on the University campus.

It was this relationship, marginal though it obviously was for him in an eventful life, that led to my being invited by Andrew Patner to join a professional physicist (Robert Wald) for a tribute to Professor Chandrasekhar broadcast by our local National Public Radio station on August 28, 1995. It was instructive to attempt to prepare myself properly for that assignment. I read a number of things by and about Mr. Chandrasekhar, including materials he had sent me over the years.

Particularly revealing were the Chandrasekhar papers in the University of Chicago Archives, which include handwritten manuscripts that are remarkably orderly. One can get a sense of the salutary discipline to which an unusually intelligent and sensitive child had been subjected to by his tutors in India some eighty years ago. One can also get a sense of how limited one's own work is by comparison.

The respect I had for Mr. Chandrasekhar was reinforced by the conversations I had, during the week before our broadcast, with a number of his scientific colleagues on the Chicago campus and elsewhere. Their ranking of him, as not far below Albert Einstein and Enrico Fermi, seemed both informed and sincere.

Perhaps most noteworthy for Mr. Chandrasekhar's fellow scientists was his ability to turn his attention every decade or so to a new field, usually a neglected field in astrophysics or physics, which he would study intensely and make his own, endowing his colleagues with a significant book on the subject. Evidently critical to his ability to do this was his prowess as a mathematician.[2] Although the Nobel Prize awarded to him in 1983 emphasized his work done a half-century before (that led eventually to astonishing and, in a sense, still unbelievable, "black holes" conjectures), it was his entire body of work which was recognized by the Nobel Prize Committee.[3]

Although most physicists are hardly likely to make much of it, the last project to which Mr. Chandrasekhar devoted himself wholeheartedly was an extended interpretation of those parts of Issac Newton's *Principia* that seemed in the direct line leading to Newton's formulation of his universal law of gravitation. This culminated in his publication of a handsome volume on that subject not long before he died: *Newton's Principia for the Common Reader.* The elegance of this book reflects splendidly the elegance of the meticulous man that we had often seen stroll the sidewalks of Hyde Park of an evening with his charming wife.

My preparation for our radio broadcast included laboring through the Chandrasekhar book on the *Principia* in which he provides for many of Newton's propositions the kind of proofs that modern physicists prefer. I had not realized, before talking to various distinguished scientists about these matters, how ill-equipped they consider themselves to be for reading Newton, something that they are not inclined to do anyway. Mr. Chandrasekhar has provided them a "translation" of Newton, embellishing it with comments of his own, many of which testify to his awe upon delving deeper and deeper into the *Principia.* His honoring of Newton in this fashion does honor to himself in turn.

The "progressive" character of modern science is reflected in the inaccessibility of Newton for practicing scientists today, however much they accept and build upon his discoveries. The isolation of modern science — a perhaps ominous isolation — is suggested by what has happened to the scientific literacy of the educated layman. In 1800 the educated layman could hope to understand some of Newton, working from his text. Two hundred years later, the educated layman can get little, if anything, from the ever-more-technical studies of nature by the most influential men of science of his own time.[4]

Somehow or other, Newton (who died in 1729) and his readers for a century thereafter could make more use than can their counterparts today of the natural human understanding of things. The geometrical mode of demonstration employed by Newton seems closer to natural things — and thus defers more to the bodily aspects of things — than the algebraic (and hence ever more "abstract" and inventive) mode employed by his successors today, however much Galileo, Newton and their colleagues tried to distance themselves from Aristotle and his truly natural understanding.[5] The contemporary, somewhat depressed, status of nature among educated folk these days does not seem to be sufficiently noticed, even though a proper grasp of nature is the basis not only of truly reliable science but also of an enduring morality among us. The modern physicist tends to be oblivious to the somewhat unintended effects of our science and its technology upon old-fashioned (if not genuine) philosophy as well as upon ordinary morality. Here, as elsewhere, we can notice the tension between the natural quest for truth, on the one hand, and the perhaps natural pursuit of justice and the common good, on the other.[6]

Even so, Mr. Chandrasekhar's pioneering effort — perhaps the most serious reading of the *Principia* by a first-rate scientist in this century — should encourage a professional physicist here and there to begin thinking about the price in genuine understanding that has been paid in order to secure the undoubted marvels of modern science and its attendant technology.

Something more of the man, and of modern science, is suggested by a conversation that I had with Mr. Chandrasekhar in the Physics Common Room before a department colloquium in April 1993. He remarked upon the fact that I was still attending the weekly Colloquium. I responded that it was like my going regularly to Orchestra Hall. He suggested that I overestimated the music I could hear at a Colloquium. It is not that I understand much of what I hear at either place, I explained, but I cannot help but admire the imagination, the competence, and the devotion I can observe in both places and occasionally I do get a glimpse of the wonderful things on display, all of which is quite instructive as well as edifying. I did not need to add what should be obvious to anyone who knows me: there is much that I have yet to learn about both music and physics.

I then asked Mr. Chandrasekhar, "I hear you are studying Newton these days. Are you finding him as interesting as you had hoped he would be?" He replied that somebody else who had heard he was studying Newton had recently asked him, "How do you feel?" And (Mr. Chandrasekhar continued) he had answered, "I am like a small boy going to the zoo for the first time and seeing a lion." There was, of course, something of the magisterial lion in Mr. Chandrasekhar as well, which encourages me to make this further observation about him. It chanced that my wife and I, upon returning home from an evening stroll on what proved to be the day of his death, noticed a very bright light in the sky. So striking was it that I called out of his house a neighbor who is on the astronomy faculty of the University of Chicago. He identified the light as Jupiter. It seemed fitting to me, upon learning the following morning of Professor Chandrasekhar's death, that this majestic heavenly display, whatever it was, should have appeared to the southwest of our house, which is where he had lived.[7]

Notes

1. The *Bhagavad Gita* in the *Mahabharata*, trans. J.A B. van Buitenen (Chicago: University of Chicago Press, 1981), p. 75. See, for the use of a passage taken from the *Mahabharata* as illustrative of the kind of "ability for unsurpassed concentration" that Issac Newton had, S. Chandrasekhar, *Newton's Principia for the Common Reader* (Oxford: Clarendon Press, 1995), pp. 258–59. I had occasion to consult with Professor Chandrasekhar about the numerology of the *Bhagavad Gita*:

> There are eighteen chapters in the *Gita*. These chapters are to be found in the *Mahabharata*, itself made up of eighteen books, which is devoted in its entirety to an epic battle that takes eighteen days ...
>
> Why eighteen? It does seem to be a significant number for Hindus ... (Eighteen *can* be seen to reflect a certain symmetry using more important numbers: five plus eight plus five ....) Subrahmanyan Chandrasekhar, of the University of Chicago, has suggested to me in conversation, that recollection of an eighteen-day battle (real or supposed) may have guided the authors of the *Mahabharata* and of the *Gita* in making their divisions. That is, he starts (in a scientific fashion?) from what might have actually been, if only accidentally. Even so, such authors might still have had to decide whether this number (as distinguished from others no doubt also available from "history") was particularly significant.

G. Anastaplo, "An Introduction to Hindu Thought: The *Bhagavad Gita*," *Great Ideas Today*, vol. 1985, pp. 264, 279 n. 30 (1985). (This is the second of seven introductions to non-Western thought that I have done for *Great Ideas Today*, an *Encyclopaedia Britannica* annual publication. I am scheduled to have a review of the Chandrasekhar Newton book published in volume 1997 of *Great Ideas Today*.) See also note 4, below.

2. See, on Lord Keynes's recognition of Newton as "a supreme mathematical technician," Chandrasekhar, *Newton's Principia for the Common Reader*, pp. 285–89. Mr. Chandrasekhar's own prowess as a mathematician and expositor is repeatedly evident in his Newton book, not least in the high praise he can confidently bestow upon Newton again and again. See, e.g., *ibid.*, p. 505: "The brevity of Newton's demonstration of this lemma in less than 15 lines of prose is to be contrasted with three pages of ours." On more than one occasion I observed the dazzling stream of mathematical statements that Mr. Chandrasekhar could write with ease on the blackboard to accompany lectures during physics and mathematics colloquia. Virtuosity in applied mathematics can sometimes make it seem that the difficulties and uncertainties, and hence both common sense and the underlying physics, in a problem have been left out of the reckoning. See, on the relation between mathematics and nature, *ibid.*, p. 268. See also G. Anastaplo, *The Constitutionalist: Notes on the First Amendment* (Dallas: Southern Methodist University Press, 1971), pp. 806–08. See, on "magic numbers" and "the shell model theory" glanced at in these closing pages of *The Constitutionalist*, R.G. Sachs, "Maria Goeppert Mayer — Two-Fold Pioneer," *Physics Today*, February 1982, pp. 46, 50–51. See, on nature, note 6, below. See also a review of a useful translation of Aristotle's *Physics* that I have prepared for the *St. John's College Review* (1996).
3. See, on black holes, S. Chandrasekhar, *The Mathematical Theory of Black Holes* (New York: Oxford University Press, 1992), p. 1:

> The black holes of nature are the most perfect macroscopic objects in the universe: the only elements in their construction are our concepts of space and time. And since the general theory of relativity permits only a single unique family of solutions for their descriptions, they are the simplest of objects as well.

See also note 5, below. In what sense are black holes objects if "the only elements in their construction are our concepts of space and time"? Does the object tend to be replaced in physics today by that which is used to make it intelligible? See, on the modern use of "concept" and of symbols, and hence the limitations as well as the power of our mathematical physics, J. Klein, *Greek Mathematical Thought and the Origin of Algebra* (New York: Dover Publications, 1968; original German text, 1934), pp. 3–4, 117–25, 163–78, 192–211; J. Klein, *Lectures and Essays* (Annapolis, Maryland: St. John's College Press, 1985), pp. 1–5, 9–10, 17–18, 27, 32–34, 43-44, 57-60, 65f, 83-84, 85f, 106, 365; L. Berns, "Rational Animal–Political Animal," in *Essays in Honor of Jacob Klein* (Annapolis, Maryland: St. John's College Press, 1976), p. 35. One can be reminded here of a remark by Sherlock Holmes quoted by Mr. Chandrasekhar in his *Newton* book (p. 454): "When you have eliminated the impossible whatever remains, *however improbable*, must be the truth." But when is it certain that adequate "elimination" has taken place? See, e.g., H. Fritzsche, "Of Things That Are Not," in J.A. Murley, W.T. Braithwaite, and R.L. Stone, eds., *Law and Philosophy: The Practice of Theory* (Athens, Ohio, Ohio University Press, 1992), vol. I, pp. 3–18. See also note 4, below. Consider, as well, the implications of the following observations:

> Since the only method available for exploring the interior of a star is the deductive one based on physical theory, it is clear that a number of assumptions will have to be made before we can arrive at a coherent picture, and that the only real test of the theory will be its self-consistency and freedom from logical contradictions. Accordingly, it is important that we be aware of the essential factors which are operative toward particular ends. The first step, then, is to get an approximate idea of the physical conditions in the interior of the stars, using a minimum of assumptions.

S. Chandrasekhar, "The Structure, the Composition, and the Source of Energy of the Stars," in J.A. Hynek, ed., *Astrophysics: A Topical Symposium* (New York: McGraw-Hill Book Co., 1951), p. 599.

4. The layman of today, upon hearing one report after another about the discovery of ever-smaller "fundamental building blocks of matter," may be moved to ask the following questions:

> Is there any reason to doubt that physicists will, if they continue as they have in the twentieth century, achieve, again and again, "decisive breakthroughs" in dividing subatomic "particles"? But what future, or genuine understanding, is there in *that*? I believe it would be fruitful for physicists — that is, for a few of the more imaginative among them — to consider seriously the nature of what we can call the "ultron." What must this ultimate particle be like (if, indeed, it is a particle and not an idea or a principle)? For is not an "ultron" implied by the endeavors of our physicists, by their recourse to more and more ingenious (and expensive) equipment and experiments? Or are we to assume an infinite regress (sometimes called progress) and no standing place or starting point? Or, to put this question still another way, what is it that permits the universe to be and to be (if it is) intelligible?

G. Anastaplo, *The Artist as Thinker: From Shakespeare to Joyce* (Athens, Ohio: Ohio University Press, 1983), pp. 252–53 (reprinting a 1974 article). Similar questions can be asked about what the limits of time or space (in any "direction") are that seem to be presupposed by physicists today. If the universe is as enormous as astronomers now say it is, is it not likely (in the light of the steadily expanded estimates we have witnessed in recent decades) to be even greater than they say, much much greater? One consequence of this conclusion, with its millions upon millions of galaxies, may be that "the Big Bang" is not a unique event of its type. Furthermore, can it be assumed (as Newton sometimes seems to do, for example) that "the centre of the system of the world is immovable"? See Chandrasekhar, *Newton's Principia for the Common Reader*, pp. 206, 376–79; Klein, *Lectures and Essays*, pp. 21, 114, 188. See, on modern science, C. Wilson, "Newton's Path to the *Principia*," *Great Ideas Today*, vol. 1985, p. 178 (1985); G. Anastaplo, *The American Moralist: On Law, Ethics, and Government* (Athens, Ohio: Ohio University Press, 1992), pp. 83–102, 620; *The Artist as Thinker*, pp. 33–941, 496.

5. There was something Aristotelian in the scope of the Chandrasekhar investigations of subjects in addition to astrophysics, mathematics, and physics. Artists of various kinds are freely drawn upon by him. See, for example, S. Chandrasekhar, *Truth and Beauty: Aesthetics and Motivations in Science* (Chicago: University of Chicago Press, 1987); S. Chandrasekhar, "The Series Paintings of Claude Monet and the Landscape of General Relativity," Inter-University Centre for Astronomy and Astrophysics Dedication Address, December 28, 1992; S. Chandrasekhar, "Newton and Michelangelo," *Current Science*, vol. 67, p. 497 (1994). He concludes his *Black Holes* book (p. 637) with this observation:

> The author had occasion to ask Henry Moore how one should view sculptures: from afar or from near by. Moore's response was that the greatest sculptures can be viewed — indeed, should be viewed — from all distances, since new aspects of beauty will be revealed at every scale. Moore cited the sculptures of Michelangelo as examples: from the excellence of their entire proportion to the graceful delicacy of the fingernails. The mathematical perfectness of the black holes of Nature is, similarly, revealed at every level by some strangeness in the proportion, in conformity of the parts to one another and to the whole.

This observation is preceded by quotations from Francis Bacon ("There is no excellent beauty that hath not some strangeness in the proportion.") and from Werner Heisenbag ("Beauty is the proper conformity of the parts to one another and to the whole."). A University of Chicago colleague recalls a Chandrasekhar observation at the faculty club: "The most important contribution of the general theory of relativity is that it shows clearly the role of aesthetic judgment in the formulation of a theory."

6. See the opening sentences of three of Aristotle's treatises: *Metaphysics*, *Nicomachean Ethics*, and *Politics*. See, on nature and morality, L. Strauss, *Natural Right and History* (Chicago: University of Chicago Press, 1953), pp. 7–8, 23, 160–61, 170f, 311–12; R.D. Masters, *Beyond Relativism: Science and Human Value* (Hanover, N.H.: University Press of New England, 1993), pp. 149–57, 235, n. 21; Klein, *Lectures and Essays*, pp. 219–39; Anastaplo, *Human Being and Citizen* (Chicago: Swallow Press, 1975), p. 328; *The American Moralist*, p. 616. See also note 3, above.

7. Johann Bernoulli, too, regarded Newton as a lion, recognizable by his "paw." See Chandrasekhar, *Newton's Principia for the Common Reader*, pp. 571–73. Others have to settle for being almost-lions, sometimes to their detriment. See *ibid.*, p. 384. Thomas Jefferson considered Napoleon Bonaparte, for example, "a lion in the field only," not in civil life. See G. Anastaplo, *The Amendments to the Constitution: A Commentary* (Baltimore: Johns Hopkins University Press, 1995), pp. 110, 331. See, on Jefferson's admiration of Bacon and Newton, *ibid.*, pp. 122, 42 and 27. Professor Chandrasekhar's own admiration of Newton culminates in the lines with which he closes his *Newton* book (p. 595):

> Ben Jonson said of Shakespeare in the First Folio: He was not of an age, but for all time! It could be equally said of Newton: He was not of an age, but for all time!

Of Subrahmanyan Chandrasekhar, in turn, it can be said that he longed to investigate not the things of an age but rather the timeless things. See the epigraph at note 1, above.

Chandra and Ramanujan

Richard Askey

I first met Chandrasekhar because of Ramanujan, but there were other sides of Chandra which I saw through the years. One deals with his constant desire to understand science. In 1983 he called and said that he had a new differential equation which he did not understand. It was more complicated than the hypergeometric differential equation studied extensively by Euler, Gauss and Riemann, and he wondered what a mathematician who knew the work on hypergeometric functions well would think when looking at this. Actually, he asked what Euler might have thought when he first started to look at the hypergeometric equation.

That was a daunting question, but he had asked if he could come to Madison for a couple of days to talk about this equation. I thought if it meant that much to him, I could not refuse. However, I had no hope of being able to see something which Chandra had missed. Chandra took a bus 150 miles to Madison, and we spent an afternoon looking at the equation. After he showed me some of his results, we computed another example. I had to admit that I did not understand what was happening. Here is what I wrote about this in the preface to [3], the proceedings of a meeting in the Black Forest in Germany held at the conference center in Oberwolfach:

> I want to close by mentioning a very interesting set of functions that I do not understand at all. These are solutions to some differential equations that arose from studying black holes. S. Chandrasekhar obtained these differential equations and solutions in a series of papers in the Proceedings of the Royal Society. A detailed treatment of them (6,000 pages or so) has been deposited at the University of Chicago, and these calculations are partly redone and summarized in his recent book [6]. Here is one quotation from this book [p. 497] that nicely 'sums up' the state of affairs. 'Nevertheless, the emergence of the various integral identities express relations whose origins are shrouded in mystery.' Chandrasekhar

> explained some of this to me, and some of the numbers that arose suggested that there were different groups that could be used to explain some of these results. These equations and their solutions should be looked at by a number of people, since something very interesting is going on.

Willard Miller of the University of Minnesota learned about this problem, and over the next few years, he and E.G. Kalnins wrote a series of papers giving the group-theoretic interpretation that was wishfully suggested in the previous paragraph.

There was another time when Chandra and Lalitha came to Madison, and then went with us to Spring Green. There is a classical theater company there which had a few remarkable actors and actresses. My wife Liz and I told Chandra and Lalitha about this company, and they were interested enough to drive up and spend a couple of days. They enjoyed the Shakespeare presentations enough to come up a second time.

However, my main contact with Chandra was because of Ramanujan. Ramanujan was a great Indian mathematician who spent five years in England, from 1914 to 1919, and then returned to India and died a year later. He had been self-educated as much as any great mathematician of the last two centuries was, and did what was thought of at that time as old-fashioned mathematics, very striking but not in the mainstream. However, through the years a number of young people who later became great mathematicians or physicists became very interested in Ramanujan's work. These included Atle Selberg and Freeman Dyson. Our view of Ramanujan is much different now.

Chandra learned about Ramanujan when he was nine years old. Here is how he described it [5]:

> It must have been a day in April 1920, when I was not quite ten years old, when my mother told me of an item in the newspaper of the day that a famous Indian mathematician, Ramanujan by name, had died the preceding day; and she told me further that Ramanujan had gone to England some years earlier, had collaborated with some famous English mathematicians, and that he had returned only very recently, and was well known internationally for what he had achieved. Though I had no idea at that time of what kind of a mathematician Ramanujan was, or indeed what scientific achievement meant, I can still recall the gladness I felt at the assurance that one brought up under circumstances similar to my own, could have achieved what I could not grasp. I am sure that

> others were equally gladdened. I hope that it is not hard for you to imagine what the example of Ramanujan could have provided for young men and women of those times, beginning to look at the world with increasingly different perceptions.

Chandra plays a number of roles in the story which follows. One of these is his finding of Mrs. Ramanujan in 1936, and obtaining a copy of Ramanujan's passport photo. However, it seems better to skip some years, to 1976, and pick up the intermediate part as needed. George Andrews is a mathematician at Pennsylvania State University who has spent his professional career doing work in areas in which Ramanujan worked. He was visiting in Madison for the academic year 1975–76, so he and I could do some joint work. In the spring, Andrews was invited to attend a meeting in Strasbourg. The airline fares at that time were a bit strange, so that traveling to Europe for more than three weeks was much cheaper than for less than three weeks. Andrews was not teaching that year, so he took his family with him to France, and then went to Cambridge to look for old papers in the Wren Library at Trinity College. Lucy Slater had suggested he look at the papers of G.N. Watson which had been deposited there. In a box from Watson's estate, Andrews came across about 100 pages written by Srinivasa Ramanujan. Anyone who had worked on Ramanujan's mathematics would have recognized the handwriting, since the *Notebooks* [13], which were in Ramanujan's handwriting, had been published in 1957 by the Tata Institue of Fundamental Research. Also, many of the formulas contained in these pages were of a type which only Ramanujan had done. Chandra's cousin, Sivaraj Ramaseshan, has a delightful article about Ramanujan [15]. It includes a description of the printers in Bombay when they were working on this job. The printers considered the printing of these notebooks one of the most exciting jobs they had ever undertaken. Ramanujan's work has had a similar effect on many mathematicians. The mathematics in these notebooks is now much more accessible, thanks to the long term scholarly work done by Bruce Berndt. See [4].

Among the formulas on the pages Andrews found were a number dealing with what Ramanujan called mock theta functions. These were functions Ramanujan discovered during the last year of his life. Ramanujan had mentioned some examples and given a general description of their properties in a letter written to G.H. Hardy about three months before he died. The words "mock theta function" do not appear on the pages Andrews found. However, Andrews had written a thesis on them and had also published some combinatorial consequences of their transformation formulas, so he immediately recognized the significance of this find.

We now know that these pages had been found by J.M. Whittaker, who had been asked to write an obituary of G.N. Watson. Whittaker visited Watson's home to look at his papers, and when he was taken to Watson's study, one of the piles of papers he looked at contained the material which Andrews later saw at Trinity's library. See Whittaker's letter to Andrews, which is reprinted in [4, page 304]. Whittaker had not worked in this area and so did not recognize the importance of his find. Robert Rankin had earlier been to Watson's home for the same reason, and while there, he had found some other material related to Ramanujan. This was the work which Watson and B.M. Wilson had done in the 1930s on the second of Ramanujan's earlier notebooks. Rankin and Mrs. Watson had arranged for the Watson and Wilson work to be placed at Trinity in the Wren library. Whittaker and Rankin decided that the new find should also be stored there. To them this material looked similar to what was already known from the early notebooks, and so did not seem that important. Andrews was the first person to understand that these sheets contain mathematics done by Ramanujan in the last 15 months of his life. Some of this is probably Ramanujan's deepest work.

Andrews studied some of the material in these pages and eventually wrote an introduction which was published in the *American Mathematical Monthly* [1]. This article was seen by John Noble Wilford, a science reporter for *The New York Times.* Wilford wrote an article on the discovery [17] of what Andrews has called "The Lost Notebook" [14]. This article was read by someone at *The Hindu*, a national newspaper based in Madras. A full page on Ramanujan appeared in *The Hindu.* Most of this page was an interview with Andrews [12], but there was also an interview with Janaki Ammal, Ramanujan's widow [10]. In this interview she said: "They said years ago a statue would be erected in honour of my husband. Where is the statue?" The writer remarked that she said this with sorrow etched all over her face.

Andrews was sent a copy of this page and he sent me a copy. It was clear that something should be done about Mrs. Ramanujan's request. It would be an error to wait for societies or governments to do something, since Mrs. Ramanujan was 81 and organizations tend to move slowly if they move at all. I contacted Paul Granlund, a sculptor whose work my wife and I knew, to ask if he would be willing to make a bust of Ramanujan. He had not heard of Ramanujan, but was fascinated by the story and said he would be willing to try. Before starting, I contacted Chandra for advice and help. I asked three questions. The first was whether this was appropriate within Indian society. The second was whether it was presumptuous on my part to try to arrange this. The third was, if this was appropriate and he felt my doing it was a good idea, would he lend the copy of Ramanujan's passport photo which he had on his office wall? G.H. Hardy had asked Chandra to try to find a photo of Ramanujan when he went back to India in 1936. Just before returning from India, Chandra was able to find Mrs. Ramanujan and she had Ramanujan's passport with the photo taken in England in 1919 before he returned to India. Chandra's brother took a photo of this picture. This is the photo we know from Hardy's book *Ramanujan* [9] and other publications where it has been reproduced. Hardy wrote to Chandra about this photo:

> I was very glad to have the photograph, which seems to me an extremely good one. He looks rather ill (and no doubt was very ill): but he looks all over the genius he was.

The full letter and Chandra's earlier one to Hardy are printed in [5]. Chandra replied that it was very appropriate in Indian society to make a bust of a great person, and he would do all he could to further this plan. Paul Granlund had said that three castings of the bust needed to be spoken for before he could start, and he would make up to ten castings if there was interest. Two problems to solve were: who would buy two of the busts and how would money for the one for Mrs. Ramanujan be raised? The first problem was solved easily; Chandra and Lalitha took one and Liz and I took another. The money for the bust for Mrs. Ramanujan could have been raised in at least two ways. One would be to find a few people who would make relatively substantial contributions and pay for it that way. The second would be to raise the money from a large number of people by asking for smaller contributions. I felt this bust would mean more to Mrs. Ramanujan if it came from a large number of people, so I wrote to friends and others who appreciated the importance of Ramanujan and said that for a twenty-five-dollar contribution they would be sent a five-by-seven-inch photograph of the bust. More than 100 people contributed, some making larger contributions than suggested, so the needed money was raised easily. The bust was made and sent to India. Chandra contacted the Indian Embassy and someone there arranged shipping via Air India. There were still problems, for the crate sat in India for a number of months without anyone seeming to know where it was. Eventually it was sent to Madras and there was a ceremony presenting it to Mrs. Ramanujan. The bust was presented by Kausalya Ramaseshan, wife of Sivaraj Ramaseshan, who at that time was President of the Indian Academy of Sciences.

I had written to a number of institutions to suggest they might be interested in buying a casting of this bust. None said they would, although Trinity College made a contribution to help pay for the bust to be given to Mrs. Ramanujan. I told Chandra about this, and he was annoyed. It is not appropriate to mention which institutions were approached, but he felt that some should have bought one. Lalitha suggested that they buy a second copy of the bust and give it to the Indian Academy of Sciences in Bangalore, where there already was a bust of Raman. Lalitha's comments at the dedication of this bust are printed in [11].

Granlund made ten castings of this bust as well as an artist's proof copy. Five of them are in India, two in England and four in the United States. Chandra and Lalitha gave their copy to the Royal Society, and his remarks on this appear in [8]. As Chandra remarked in [8], the President of the Royal Society, Sir Michael Atiyah, said that the bust would be installed in the President's office as a companion to the bust of Dirac already there.

The other copy in England came about in the following way. After Chandra and Lalitha gave one to the Indian Academy of Sciences, I received a substantial check from the Raman Research Institute Trust to help pay for the bust given to Mrs. Ramanujan. This, along with the contributions of individuals, meant that there was extra money in the Ramanujan Bust Fund. At that time there was no copy of the bust in England, where Ramanujan had spent five years. There should have been, so I wrote Robert Rankin and offered to pay half of the cost of the bust if the other half could be raised. He wrote to some institutions and individuals, and a general plea for contributions was published in the Newsletter of the London Mathematical Society. They raised two thirds of the cost. Rankin had a beautiful plinth made in Glasgow, and the bust is now in the Pure Maths Library in Cambridge. My wife and I took this bust to England. While there we met Bela and Gabriella Bollobas. She had done a bust of Littlewood, which at that time was in the Combination Room at Trinity, where fellows would retire after dinner for port, sherry, cheese, fruit and conversation. She mentioned that she had wanted to do busts of Hardy and Ramanujan as well, but the Granlund bust did this for Ramanujan, so Hardy's was the only one missing. I offered to pay for the casting of a bust of Hardy if she made one. This seemed an appropriate use of the remaining money given for the Ramanujan bust. She did, and now there are busts of Hardy, Littlewood and Ramanujan at Trinity.

To summarize Chandra's role in the story of Ramanujan's bust, he found Mrs. Ramanujan and had a copy of the passport photo made. In his biography of Chandra [16], K.C. Wali quoted Chandra, paraphrasing Hardy, as claiming that his discovery of the passport photo was one of his most important discoveries. Lalitha quoted Chandra as saying, "To this day that is my best contribution to mathematics." For me, his saying that a bust of Ramanujan would be a very good idea and that he would do all he could to help was just the push I needed to carry it through. The contribution from the Raman Research Trust, which was made after the second bust the Chandrasekhars bought had been sent to India, made it possible for me to offer a lower price for a bust to the British mathematical community. One of the busts which was acquired by an Indian institution was bought because Chandra wrote to a friend there and suggested that they should have one. The others in India were obtained after people there saw or read about the one Chandra and Lalitha had placed in Bangalore.

The fact that Rankin was able to raise more money than needed made it possible for me to use the remaining money to pay for the casting of the Hardy bust. I am grateful for what Lalitha and Chandra did and for all the others who helped in this project. My payment came when I met Mrs. Ramanujan in her home and saw how she cherished the bust. She draped it with a garland every day. Liz and I have spent a number of very pleasant times with Chandra and Lalitha in Chicago. One was when we delivered their copy of the bust. They invited us to stay overnight, and the next morning, when we arose, there was Chandra sitting looking at the bust. He said that it was remarkable, for if you project it back to two dimensions you get the photograph which had hung on his office wall for decades. I still remember his look of contentment.

In the fall of 1993, I went to a meeting in Chicago and had dinner with Chandra and Lalitha at the Quadrangle Club. Chandra told me about his favorite bookstore in Chicago, the Seminary Co-op. A year later Liz and I were in Chicago for another meeting and had lunch with Chandra and Lalitha at the Quadrangle Club again. Chandra remembered that he had told me about the Seminary Co-op Book Store and asked if I had gone there. I had to admit that I had not. After lunch they walked us to the Chicago Theological Seminary and took us to the stairs which led to the book store in the basement. We went down and saw why he was so insistent on our going there.

It is a marvelous bookstore, and he wanted to make sure we did not miss it. This was typical of Chandra. In his quiet way he made sure that you did not miss some of the good things in life. He did this directly at times, and regularly through his writing. What he really did is show us how a great life can be led.

Chandra closed his talk on Ramanujan [7] as follows:

> My own view, sixty-six years after my first knowing of his name, is that India and the Indian scientific community were exceptionally fortunate in having before them the example of Ramanujan. It is hopeless to try to emulate him. But he was there even as the Everest is there.

While Chandra would be the first to protest if his name were used in the same way as he used Ramanujan's name, I feel that we have been very fortunate to have had Chandra among us for many years. While he set an example which we cannot hope to match, it is one which leads us to ask more of ourselves than we thought possible.

Notes

1. G.E. Andrews, "An introduction to Ramanujan's 'lost' notebook," *American Mathematical Monthly* 86 (1979) 89–108.
2. G.E. Andrews *et al.*, *Ramanujan Revisited* (Academic Press, Boston, 1988).
3. R.A. Askey, T.H. Koornwinder, and W. Schempp (eds.), *Special Functions: Group Theoretical Aspects and Application* (Kluwer, Dordrecht, 1984).
4. B.C. Berndt, *Ramanujan's Notebooks*, Parts I, II, III, IV (Springer, New York, 1985, 1989, 1991, 1994).
5. B.C. Berndt and R.A. Rankin, *Ramanujan: Letters and Commentary, History of Mathematics*, Vol. 9, Amer. Math. Soc. and London Math. Soc., 1995.
6. S. Chandrasekhar, *The Mathematical Theory of Black Holes* (Oxford University Press, Oxford, 1983).
7. S. Chandrasekhar, "On Ramanujan," in [2], pp. 1–6.
8. S. Chandrasekhar, "Reminiscences and discoveries: on Ramanujan's bust," *Notes Rec. Royal Soc. London* 49 (1) (1995) 153–157.
9. G.H. Hardy, *Ramanujan: Twelve Lectures on Subjects Suggested by His Life and Work* (Cambridge University Press, Cambridge, 1940).
10. *The Hindu*, June 21, 1981, "His papers disappeared mysteriously," "Interview with Janaki Ammal."
11. Indian Academy of Science, "Presentation of the bust of Srinivasa Ramanujan," Patrika, Newsletter of the Indian Academy of Sciences, April 1985, No. 10. Reprinted in *Current Science* 59 (1990) 1316–1318.
12. N. Ram, "Ramanujan's last wıll and testament," *The Hindu*, June 21, 1981.

13. S. Ramanujan, *Notebooks of S. Ramanujan*, Vols. I and II (Tata Institute of Fundamental Research, Bombay, 1957).
14. S. Ramanujan, *The Lost Notebook and Other Unpublished Papers* (Narosa Publishing House, New Delhi, 1988). Also available from Springer-Verlag, New York.
15. S. Ramaseshan, "Srinivasa Ramanujan," *Current Science* 59 (1990) 1309–1316.
16. K.C. Wali, *Chandra: A Biography of S. Chandrasekhar* (University of Chicago Press, Chicago, 1991).
17. John Noble Wilford, "Mathematician's final equations praised," *New York Times*, June 9, 1981.

Reminiscences of Chandra

James W. Cronin

During the past 20 years my wife and I became rather close friends with Chandra and his wife Lalitha. During much of this time we have lived in the same apartment building on Dorchester Ave. Chandra and Lalitha would visit for luncheon or tea rather frequently and would have discussions ranging over many topics, but principally concerning science, its philosophy and its personalities. Chandra was the only truly great scientist (among a handful in this century) that I had the privilege to know well.

I first encountered Chandra as a student in one of his classes given at the University of Chicago in 1954. The subject was magneto-hydrodynamics and Chandra was describing experiments he was supervising. The experiment consisted of a study of convective motions in a rotating pool of mercury placed in a magnetic field with a thermal gradient in the direction of the magnetic field. I was overwhelmed as Chandra effortlessly developed and then proceeded to solve the complicated partial differential equations which described the experimental process. I remember almost none of the details of the class. But the impression of Chandra as a brilliant scientist was vivid and lasting.

When I returned to the University in 1971, Chandra and I gradually began a friendship. We would often visit one another's office to chat. Chandra had a rather formal way if one did not know him well. Often in conversation his colleagues might defer to his statements and opinions without contesting them. Early in our encounters I decided to tell Chandra what I really thought about a particular matter even if I disagreed. I believe Chandra appreciated my candor and found it missing in so many of his interactions with others.

Growing out of our early chats was a desire on my part to learn something about general relativity. When Chandra offered such a course in the fall of 1980 I decided to follow the lectures. I was determined not to be absent. Early October, 14 I was awakened by the news that Val Fitch and I had been awarded the Nobel Prize. Even on this day I was determined to attend the class. I postponed all interviews and insisted that no press people be allowed to know my whereabouts. When Chandra entered he came up to me to offer congratulations and then proceeded with his lecture. My attendance on that special day served to cement what was already becoming a warm relationship.

I gave Chandra a draft copy of my Nobel lecture for comment and criticism. His principal remark was that in the introduction I was too deferential to the Nobel Foundation, being effusive about the great honor they were bestowing on me. Chandra said it should be just the opposite. We were honoring the Foundation for accepting the Prize! In 1983 when Chandra was awarded the Nobel Prize I had my revenge as I found he had ignored his own advice in the introduction to his Nobel lecture.

Chandra would often come to my office wanting to discuss a new scientific biography or autobiography. He deplored pomposity of scientists who reveled in the accomplishments of their younger days. For Chandra a life in science was one dedicated to the understanding of nature. Personal recognition and personal competition should play no role. Chandra and I had endless disagreements on this subject. I maintained that there was no correlation between character and success in science. A brilliant scientist could indeed be a scoundrel. I believe that subconsciously Chandra held himself as a reference to which others might be compared.

Chandra's extraordinary character was displayed in his famous dispute with Eddington. He never spoke publicly against Eddington, while having been publicly humiliated by him. Chandra spoke often to me about this affair. He spoke of his effort to get support for his side of the dispute by sending his paper for comment by the most distinguished scientists of the day. None spoke up in his defense. It was clear to me that this incident had a profound effect in Chandra's scientific career, leading him to subjects of scientific inquiry that were remote from the problem of stellar collapse. It led him to investigate aspects of astronomy that at a given time were not in the mainstream of attention-getting research. The result was a series of monographs which have become classics of astronomy.

During this period Chandra resumed the area of science that he avoided early in his career and wrote the "Mathematical Theory of Black Holes." His style of working was very disciplined. He never spoke in terms of scientific work bringing him pleasure. His motivation seemed to be one of a lifelong dedication to science. There was a touch of melancholy in his demeanor. He lamented that he had few colleagues at Chicago who appreciated his work. On numerous occasions he would fly for a weekend to Oxford, where he would spend a day discussing his work with Roger Penrose. However, when a new insight appeared he became very excited and would give a seminar on the subject. He would also come to my office to talk about a new discovery. I remember his discussing with great enthusiasm the scattering of gravitational and electromagnetic waves as a coupled system. He must have realized that it was difficult for me to fully understand the details, but he enjoyed having an interested listener.

Chandra gave many lectures. Normally he would read them, but occasionally he would depart from the text to explain a figure or make a sketch on the blackboard. The freshness and spontaneity of these diversions were wonderful. I knew Chandra well enough to suggest that he speak in this manner more often, but he did not heed my advice. The written texts of his lectures are eloquent, often with references to literature and art. Chandra would always give me copies of his papers and lectures. Reading any one of these conveys an insight into the particular subject that is profound.

Chandra stated to me that after the completion of the "Mathematical Theory of Black Holes" he would no longer continue his scientific work. I really did not believe this. Scientific work was the essence of Chandra's life. Being asked in 1987 to give lectures on Newton stimulated Chandra's last work. In our conversations we always discussed what was on Chandra's mind. As he continued to study Newton, Chandra's evaluation of Newton became one of awe. Chandra would independently prove some proposition of Newton and find that Newton's proof was always more economical and more eloquent. Chandra was exhilarated by every page of the Principia and spoke of the sense of "entering into Newton's mind." He could not refrain from comparisons of Newton with other scientists. How could he have written the entire Principia in such a short time? After several years of work on Newton, Chandra by sheer determination wrote his last book while in failing health.

During the last year we were together with Chandra and Lalitha several times. The last occasion was a luncheon with the Chandrasekhars and the Sterns. Dick Stern is a novelist and a professor of English. Chandra was in fine spirits, regaling us with stories of great humor about events and personalities in the Cavendish laboratory in the 30's. Relaxed and at ease, Chandra was a delightful guest. His passing left an emptiness that can only be compared with the passing of a member of one's family.

Some Personal Recollections of S. Chandrasekhar at Chicago and Oxford

R.H. Dalitz

Introduction

In 1956, I became a colleague of Chandra in the Physics Department and the Enrico Fermi Institute for Nuclear Studies (EFINS) of the University of Chicago. Owing to visa delays, due to the smallness of the annual quota for the Australian-born, I was not physically at Chicago until March of 1957. At that time the Chandrasekhars were still based at Williams Bay, Wisconsin, near the Yerkes Observatory. However, they were also renting a small apartment in the Hyde Park district where the University of Chicago was located, since Chandra had begun his practice of giving his lectures on the Campus on two (or three) days of the week, driving down early one day and back the next day (or the day after that), staying one (or two) night in their Hyde Park apartment. His time at the University was a busy period for him and I can recall no contact with him in my first years at Chicago, although I had met him briefly and seen him frequently in the distance.

A Call from Yerkes

One morning in February 1960, I had a call from Chandra. He seemed quite excited and I heard him telling me that I had been elected to the Royal Society. I was astonished at this news and said so. I knew, of course, that I had been a candidate for election but I had little reason to expect any outcome from this. I was also surprised that Chandra had wanted to let me know about this in advance, to be the first person to inform me.

To explain the significance of his action, I should make a few remarks. In common with the procedures of many societies, the Royal Society has a number of sectional committees, each covering a number of areas of research and charged with the selection of a short list of the best candidates in their area from a long list proposed by the Fellowship. Council would then deliberate over these short lists and settle on their overall final list, which is then sent out in February to all the Fellows, allowing them a month to consider this final list. In principle, it is possible for the Fellows to make objections to candidates on this final list, if they see good reason for doing so, and this has happened occasionally. To avoid the possibility of embarrassment to any candidate whose name may be on Council's list but not on the Society's final list, which is made public in the middle of March, the Fellows are enjoined to secrecy about the names on Council's list when they receive it in the middle of February. It is this law which Chandra had violated!

However, that's what Chandra did, and it forged a link between us. Previously I had supposed him to be a fairly formal person, unlikely to buck the rules, but not so now. Later I came to see him as a man with a very strong will, but a will over which he had full control.

A Knock on My Door

One afternoon, some time about 1960, Chandra knocked on my office door. He came in saying that he had a problem about which I might be able to give him some help. I don't recall precisely what this problem was, except that there were two fluid separated by their common surface, at least one of the fluids being viscous. The boundary conditions to be met at this surface were not at all the usual ones but were part of the problem. I realized that we would know what these conditions were if there were a third layer between the two fluids with thickness, say d. This meant examining a much more complicated system, of course, and then taking the limit $d \to 0$, to reach Chandra's case. I recalled that I had once solved a not-dissimilar problem, when I was working in aerodynamics, a long time ago. I was not able to take Chandra's problem to the end. As we stood by the blackboard, Chandra said, "That's an interesting suggestion. It gives me something to think about." He gave me thanks and went on his way, leaving me much perplexed; why should Chandra be needing me to tell him how to solve such a problem, entirely in his area of expertise and not at all in mine these days? Well, I had plenty to do and I turned back to get on with my tasks.

The explanation of this small episode came about a week later. During World War II, Chandra had been doing work for the U.S. Government at their Ballistic Research Laboratory at the Aberdeen Proving Grounds in the state of Maryland, having been recruited by J. von Neumann, who was then a consultant there. Chandra worked there half-time, three weeks in Maryland and then three weeks at Yerkes, and so on. He had British nationality, so it took some time for him to receive a clearance for doing this work, but there were many other British scientists already involved in U.S. war-related projects. He began this work early in 1943, and continued it up to the end of the war. At some time after the war, many of the scientists with experience of war-related work felt that they might make themselves available to the U.S. Government for making quantitative estimates on questions in many areas of national or international concern. Chandra was a member of such a group, which could not meet very often. Its members felt that it would be helpful for their projects if each isolated member could recruit another scientist nearby, with whom he could discuss the problems he was working on. Chandra was proposing that I should work with him in this way. So, Chandra had given me an oral examination and I was now learning that I had passed. In the end, I did not take up this offer; I had a heavy load of students to get through their Ph.D.'s and I needed to be able to spend more time with my family. It wasn't really possible for me to join.

In retrospect, I have sometimes wondered whether it might have been a good idea to accept Chandra's proposal. I could have learned a great deal, and quite quickly, from close work with Chandra.

The Astrophysical Journal

The Astrophysical Journal was initially a journal owned by the University of Chicago, but due largely to Chandra's efforts it became at the same time the national journal of the American Astronomical Society (AAS) by the end of 1951. Having brought these two bodies together, Chandra found himself compelled in 1952 to accept the managing editorship of the journal, in order to ensure the success of this arrangement. He took this as a challenge and soon made *ApJ* the outstanding journal for astrophysics world-wide. Quality was his criterion and he became an "autocrat," the complete master, and totally responsible for the journal. He limited his work as editor for this journal to quite definite periods each week. He acted as referee for a large proportion (perhaps three quarters) of the papers submitted. When he did send a paper to a referee, it usually arrived there with comments and a list of questions. The journal was printed by the University of Chicago Press, a block away from his office, and for papers which were most important, in his view, he would often correct the proofs himself, then walking to the press with them, since this got the paper into *The Astrophysics Journal* at least an issue faster than mailing it back to the author for corrections.

It was a marvel that Chandra was able to carry the burden of his extended editorship, along with his lectures and his research. This was given up in 1971, after a year's trial for his replacement.

Chandra's First Postwar Sabbatical

All through his editorship of *The Astrophysical Journal*, sabbatical leave was not possible for the Chandrasekhars. In 1971, they made plans for a sabbatical, to be spent at Oxford. He came as a Visiting Fellow of All Souls College, where Dennis Sciama and I were regular Fellows, for the Hilary and Trinity terms of 1972. They lived in a College flat, one of a block of flats All Souls had built on Iffley Turn to house their Visiting Fellows. Iffley is a very green place, with many trees, and situated on a hill-slope running down, past the Iffley church built in Saxon times, down to the river Thames. It is a very pleasant place and they were very happy there.

Chandra took part in the usual Oxford life for scientists, giving for example the Halley Lecture on Astronomy and Terrestrial Magnetism entitled "The Increasing Role of General Relativity in Astronomy" on May 2nd, and other seminar talks, at Oxford and elsewhere, and attending those by others. He also finished paper III, "Vacuum Metrics and Carter's Theorem," of his series of papers "On the Stability of Axisymmetric Systems to Axisymmetric Perturbations in General Relativity," and sent it off for publication in *ApJ.* He was instrumental in getting Roger Penrose appointed to the Rouse Ball Professorship at Oxford. (Penrose had previously held a Chair at Birbeck College, a small college in London specializing in evening classes for part-time students.) Chandra was awarded an honorary D.Sc. degree on Encenia Day, June 21st. He received the degree from the Chancellor in the Sheldonian Theater, and walked in the procession of the university officials, the heads of Houses and the honoured graduands from the Sheldonian to the back gate of All Souls and thence, with Lalitha, into the vast Codrington Library, where luncheon was prepared for about 250 persons. On a sunny day, these ceremonies all make for some most impressive and colourful sights. It would be interesting to give here the citation for Chandrasekhar, since such citations at Oxford often have a humorous ring to them, but it is not practical for me to do this here, since the citation was in Latin.

Classical General Relativity

This was the last area of physics which Chandra made his own, beginning in the 1980s and following the publication of his book *The Mathematical Theory of Black Holes.* He was well pleased with this work — as he said to me several times: "There are so many new things being learned in this field, and we have no competition; where are all the others?" — problems were being solved exactly, showing interesting features, especially on the interaction between gravitational waves and black holes. Their physical features showed parallels with phenomena known in other areas of physics. For example, the scattering amplitude obtained has the Breit–Wigner form, a form well known in photon–nucleon, pion–nucleon, and nucleon–nucleus processes, to mention a few cases. This parallelism is not really surprising because these properties all arise from very general conservation laws, which have their effect in similar ways, irrespective of the type of particle under consideration. The orders of magnitude for gravitational wave scattering are of course quite different from those for (say) pion–nucleon scattering.

Chandra arranged a set of coherent publications in *Phil. Trans. R. Soc. Lond. A*, to draw attention to this open situation. The issue dated 15 September 1992, No. 1658 of Vol. A340, consists of only five papers, all on topics from classical general relativity, namely: separation of variables, symmetries of stellar models, numerical solutions, rapidly rotating stars, and a scattering approach to non-radial oscillations of stars, none of them authored by Chandra. He had commissioned these papers to show physicists the wide array of problems available, and being solved, today. Of course, the proceedings of many Royal Society Discussion Meetings have quite often been published in a single issue. The new feature of this *Phil. Trans. R. Soc.* issue was the use of this format without any meeting, a precedent worth noting.

Ramanujan's Bust

In 1936, near the end of Chandra's first return visit to India, during which he married Lalitha, he searched for Mrs. Ramanujan, finding her in Triplicane, a southern suburb of Madras. His mathematician colleague G.H. Hardy was preparing his book about Ramanujan and had asked Chandra to find a suitable photograph of him to go with this book. The photograph in Ramanujan's passport turned out to be in excellent condition, so Chandra had a copy made.

George Andrew's discovery of Ramanujan's *Lost Notebook* in 1976 led to interest by several major newspapers, and ultimately to an interview of

Mrs. Ramanujan by the paper *Hindu*, in which she had complained that she had never received a statue of Ramanujan, although this had been promised her. When Richard Askey heard of this, he sought funds from international mathematical societies and private persons to make a bust. A sculptor at Gustavus College, St. Peter (Minnesota), Paul T. Grunlund, was willing to try to make such a bust, working only from the passport photograph, but only if four copies were made, each at the price of $3,000. One was given to Mrs. Ramanujan in 1983, one was taken by Askey and the two others by Chandra and Lalitha, one of which they presented to the Indian Academy of Sciences on the occasion of its Golden Jubilee Celebrations at Bangalore in February 1985. It is now housed in the Raman Research Institute at Bangalore. It appears that the sculptor also made one for himself, which is presumably the copy known to be on show in the Mathematics Department of Gustavus College.

The 50th anniversary of Chandra's election as F.R.S. came in 1994, so the President of the Royal Society (Sir Michael Atiyah) arranged a small dinner party to mark this occasion. In the course of this dinner, after several speeches had been given about Chandra's longevity and achievements, it was revealed to us that Chandra had a present for the Royal Society, the second of their two Ramanujan busts. Chandra made a speech about the history of the Ramanujan busts, which was later printed in *Notes and Records of the Royal Society* (1995), and our president formally accepted the bust on behalf of the Society. [In a footnote to Chandra's speech, it is stated that five more copies were made later. We know that one of these is in the Library of the Faculty of Mathematics at Cambridge University, another is at the Tata Institute for Fundamental Research (TIFR), at Mumbai, and we have learned from Dr. Askey (via Dr. K. Srinivasa Rao at the Institute of Mathematical Sciences, Madras, known as MATSCIENCE) that a third bust is held by the Defence Department, at New Delhi. We understand that the fourth bust is with Andrews, and the fifth is held by the Vaughn Foundation of Texas.

A different bust was made in Madras by a local sculptor named Masilamani, and was unveiled in March 1993. This now stands in the entrance of the Ramanujan Institute for Advanced Study in Mathematics, University of Madras. No copies of it are known.]

A Return Visit to Chicago

In July 1989, I visited the University of Chicago on the way home from a conference in Vancouver. I spoke in EFINS about the many advantages the reactions $pp \to K^0 p\pi^+$ and $K^0 p\pi^-$ offered for the analysis of the neutral kaon decay processes, since experiments on these processes were shortly to begin with LEAR at CERN. As I began to speak, I noticed Chandra come in to sit at the back of the room, something which I'd never seen, nor even expected to see, before, since my topics are generally outside his area of experience. As I left the seminar room, I saw a notice about a seminar by Chandra on the next day, so I decided that I would attend his seminar.

The next morning, Chandra took me down into the Special Collections Department of the Regenstein Library of the University of Chicago to show me his extensive archives there. He particularly wanted to show me the notes he had taken of a lecture course given by Dirac in the middle 1930s on the subject of spinors. This was a lecture course which Dirac gave from time to time, on no regular timetable. Besides Dirac's book on spinors in Hilbert space, and Chandra's notes, there exist notes from two other courses he gave on spinors, both at Cambridge, one in 1940/41 and the other in 1957. The book is for n-dimensions; the lectures are for four-dimensional space–time.

At lunch, Chandra spoke to me at some length concerning his feelings about his books. Each was a work of art, like a painting, he said — something coherent, not to be changed. Indeed, when a publisher wished to publish one of his books, Chandra would stipulate that the book should not be refereed. He wasn't interested in the views of the critics, he said — who was his book for, after all? It was for *himself*, expressing his views of the things. It was not for the purpose of communication — except for putting out his view. I agreed with him that he should not be *required* to change his books in any respect.

Chandra's seminar audience were an informal lot, consisting of his research associates and Fellows; also some students, but not his own, since he was no longer taking research students by that time. I was surprised to hear Chandra take issue with a post-doc who asked a question, which Chandra thought he should answer for himself. Chandra refused to answer the question, saying rather crossly: "I recall when you asked that question before and I told you what paper to read. Did you not do as I asked you to? Apparently not!" and then going on with his talk. It was an excellent seminar, of course.

Chandra and Paul Dirac

Chandra arrived in Cambridge (England) in October 1930. At once, he began to attend the lectures on quantum mechanics given by P.A.M. Dirac through the Christmas and Lent terms of 1930/31. He was very impressed by these lectures and by Dirac's personality. He told me once that he had attended the full course of Dirac's lectures three times, in his years at Cambridge. In later life, when he met Dirac at a conference, he happened to mention this to him. Dirac was astonished and asked: "Why did you do that?" Chandra replied: "If I had told you that I had listened to the same Beethoven concerto on three occasions, you would not have found that astonishing."

However, Chandra was very critical of Dirac's statement, recorded on a blackboard in Moscow University, but repeated many times:

"Physical laws must have mathematical beauty."

In a paper written about 1985, then circulated privately, and finally presented at Telegdi's 60th birthday meeting at CERN on 11 January 1987, and published in its Festschrift ("Festi-Val") in 1988, Chandra said that it is not the mathematical form of general relativity which is so beautiful, but the physical idea which it expresses, the equality of the inertial and gravitational masses. He emphasized that the mathematical forms of physical laws are not immutable. They may change when we explore phenomena in new physical circumstances. Over sufficiently small regions, the general theory of relativity reduces to the special theory of relativity. No doubt, he said, in times far in the future, the equations of general relativity may be extended to become equations of some super-theory, but we have no clear indications of this yet.

Beauty in Physics

Chandra expressed his thoughts on beauty in science quite early, as we can see from his book "Truth and Beauty" (1987), which reprints three such talks. On 2 April 1979, he spoke on "Beauty and the Quest for Beauty in Science" at the International Symposium R.R. Wilson at Fermilab (Batavia, Illinois), a topic particularly appropriate in view of Wilson's firmly expressed artistic style as Director of Fermilab and elsewhere. His inaugural address on 6 February 1985 at the Golden Jubilee Celebrations of the Indian Academy of Science was entitled "The Pursuit of Science: Its Motivations" and explored the motivations and the style of a dozen or so of the outstanding physicists, ranging from Keppler to Dirac, noting their variety and their commonly expressed pleasure in making a discovery which drew together and illuminated previous knowledge. At Hamburg on 18 September 1986, in memory of Karl Schwarzschild, he spoke on "The Aesthetic Base of the General Relativity," extolling the simple and elegant virtues of the Kerr metric for rotating black holes and its expression in terms of the Ernst equations, and pointing out the remarkable fact that the calculations to describe the collision of two plane, thin, impulsive gravitational waves lead to the same Ernst equations. Thus the use of the Ernst equation leads to a recognition of a structural unity in the solutions for these widely different physical situations. By this time, Chandra was not so far from Dirac, in the latter's criterion for beauty in physics.

Later, on 10 May 1989, he lectured to the annual meeting of the Am. Acad. Arts and Science on "The Perception of Beauty and the Pursuit of Science," collecting together statements by a number of mathematicians and theoretical physicists, several poets and one playwright, engaging the reader's thoughts on their relationships without reaching any conclusion. In 1992, Chandra, as the outstanding Indian astrophysicist, was asked to give the inaugural lecture at the dedication ceremonies for the Inter-University Center for Astronomy and Astrophysics (IUCAA). He did so on December 28th, under the title "The Series Paintings of Claude Monet and the Landscape of General Relativity," his thought being that different gravitational phenomena led back to the same inner structure of general relativity, that provided by the Ernst equations, just as Monet painted various views of several trees, from different directions, at different times of the day, at different seasons of the year, and so on, but always being provided by the same core, the several trees themselves.

Newton's Principia

In July 1985, the organisers (S.W. Hawking and W. Israel) of the Newton Tercentenary Conference, held in Cambridge (U.K.) on 29 June–4 July, 1987, asked Chandra to contribute "an article on Newton and gravitational theory after Newton," as an appropriate opening address for the conference. However, the organisers also set the deadline for the receipt of the manuscripts as 31 August 1986, since they wished to have the tercentenary conference book, *300 Years of Gravitation*, in the hands of the participants as the conference began. Chandra accepted their invitation and took their charge very seriously. He decided that he would select some of Newton's propositions, use modern techniques to give proofs of them, and then to compare his proofs with those of Newton.

In February 1986, Chandra wrote to say that his progress had been very slow and that he would not be able to provide a satisfactory manuscript for the conference book on the topic proposed in the time available. Instead, he proposed the topic "The Aesthetic Base of General Relativity." In July, he wrote saying that this article would be inappropriate to the general theme of the book, and that it should therefore not be included in the book; in the meantime, he hoped to be able to carry out his intentions of studying the Principia, at least in some measure, by the date of the conference. As mentioned earlier, Chandra gave a lecture, with essentially the same title as proposed in his February letter, at a memorial meeting at Hamburg for K. Schwarzschild in September 1986; a week later, presumably close to a delayed deadline, he wrote to Israel to confirm that he definitely did not want this article to be printed in the book, even if it were modified to make it appropriate for the Newton Tercentenary rather than for Schwarzschild's memorial.

We know that, when Chandra got deeply into the Principia, he was profoundly moved by the elegance and directness of the geometrical proofs given by Newton, qualities which went far beyond his expectation. These proofs implied the knowledge and experience of a body of geometrical and other mathematical relationships, which Newton could draw upon when needed, but which Chandra did not have readily at hand, as part of his intuition. Newton's proofs were not as sophisticated as those with modern methods, but they were most economical and highly original, devised for Newton's specific purpose. He found Newton's treatments admirable and it was not long before he felt it urgent to make his conclusions known and to have other physicists join him in his admiration of Newton.

It has been difficult to determine what lecture Chandra gave as the introductory address at the Newton Tercentenary Conference. The book *300 Years of Gravitation* gives us no evidence about its content,[1] since Chandra's talk is missing, being replaced by four pages of introductory remarks about Newton and the importance of his Principia, written by Hawking. This is consistent with the correspondence quoted above, since it implies that neither of the possible lectures mentioned by Chandra could appear in the book of the conference. The recollections of those who attended the conference vary widely, from one extreme to the other, but those with the firmest recollections said that it had a part about the Principia and a part about general relativity, but disagreed on their order in the talk. We were unable to find a copy of the conference programme, nine years after the event. Finally, the only contemporary evidence we were able to find was that of the notes taken by Professor D.N. Page at the lecture. We will quote these notes, with the author's permission:

" 'S. Chandrasekhar: The Aesthetic Base of the General Theory of Relativity,' December 1684 – May 1686, the Principia was conceived and written in Trinity. I decided to try to read it, but that I estimated would take a year, so I took 15 propositions, tried to prove them on my own, and then compare them with Newton's proofs. I was astounded at his originality and lack of superfluity.

Dirac said general relativity has a simplicity and elegance all its own. How does its aesthetic relate to its applicability?

Efforts have been made to test GR deviations from Newton theory: rates of clocks, light deflection, perihelion precession, and the slowing of light near a massive body. But no tests have been made in strong fields, nor do any seem forthcoming in the foreseeable future. A confirmation of GR would be if a prediction of that theory, and that theory alone, were observed. Black holes would not be a confirmation unless the metric around them could be measured. There is no solid ground for GR, so on what do we build our trust? On the aesthetic beauty and consistency of the theory."

At this point, Chandra's talk moved over to general relativity, close to the "The Aesthetic ..." talk he had given at Hamburg almost a year before. His talk ended with the following thoughts: "Einstein: 'Anyone who comprehends the theory must see its magic.' I see it in its mathematical structure." Almost four months later, Chandra gave a two-hour-long public lecture at a symposium to celebrate the tercentenary of "Isaac Newton's *Principia Mathematica* (1687–1987)" at the University of Alberta, Edmonton, which showed how his admiration for Newton and the methods he used had continued to increase as his study of the Principia progressed. When he was asked to speak, on 22 December 1987, the centennial of Ramanujan's birthday, at the Ramanujan Centenary Conference, organised by MATSCIENCE and the National Board for Higher Mathematics in India, Chandra chose to speak about the outstanding elegance of Newton's Principia and to marvel at Newton.

Ten Lectures on the Principia, at Oxford

It became apparent to us that Chandra would be glad of an opportunity to give a series of lectures on Newton's Principia at Oxford, in the Trinity term of 1991. It happened that Roger Penrose had already arranged to be away from Oxford during that term, on sabbatical leave, but that I would be there. A notice was sent out widely, to the effect that Professor S. Chandrasekhar would lecture on "Newton's Principia: The Intellectual Achievement That It Is," over five weeks, two lectures each week (Monday and Tuesday), starting April 22nd. By Oxford standards, the lectures were well attended, the regular audience being about sixteen persons, most of them being established historians of science and half of those being already well-informed about the Principia. Almost all of them attended all of the lectures.

We were fortunate in being able to house Chandra and Lalitha in an All Souls College flat at Iffley, in the same block as the flat they had occupied during Chandra's Visiting Fellowship. It was wonderful weather during their five weeks at Oxford and Iffley was at its freshest and greenest. It was a happy and relaxed time for them both. Chandra was free, apart from giving his lectures, which was his wish.

After the tercentenary conference in 1987, having noticed that Chandra's lecture was missing from the conference book, the Science Editor of the Oxford University Press enquired of Chandra whether he would be willing to have O.U.P. publish his lecture. He replied that he did want to publish his remarks on the Principia, but only after much more study and in a much more extended form. While he was in Oxford in 1991, he made contact again with O.U.P. about his project of publishing a book on the Principia, based on his ten 1991 lectures but going far beyond them. The O.U.P. response was very positive; in a letter dated 10 May 1991, O.U.P. confirmed their interest in publishing Chandra's book and their willingness to accommodate Chandra's wishes concerning the layout of the book. A contract was signed early in 1992.

However, Chandra scorned the historians of science and their questions of detail. His purpose was rather to show them the beauty of Newton's constructions and so lead them to appreciate and admire the man and his book. He simply had no interest in their questions. At one lecture, the audience asked him to stay with them after the lecture and to take part in a general discussion about a number of outstanding questions they had about its text and its time relationship with other work Newton may have done and used without acknowledging the source (possibly, unpublished manuscripts of his own), questions which the specialists on Newton and all his works had not been able to resolve. There was no meeting of minds on the questions raised and no progress was made in this discussion. Nevertheless the audience continued to attend; they certainly wished to hear what Chandra had to say, even if it did not solve any of their own problems with the Principia.

Newton's Principia for the Common Reader

Chandra's book was published on 15 June 1995; the records show that one pre-publication copy went out to him on May 12th. The book is elegant, being well printed in the O.U.P. style. It is a book which will arouse controversy and so bring much new research to bear on Newton and his Principia. This was Chandra's ultimate book, marking the completion of his professional life. Beyond it, he appeared to have no clear goals.

Almost all the reviews of this book have remarked on the fact that it does have many errors, something very unlike any other book by Chandra. Probably this was the result of haste, for he felt that he was racing against time, determined to complete the text so that it would be in the hands of the printers before his end came.

Conclusion

I saw him for the last time at Oxford, in mid-March 1995. I took Chandra and Lalitha out to see Rudolf Peierls, who was a contemporary of Chandra's at Cambridge. Peierls had moved out of Oxford in 1994 to Oakenholt, an estate in the country district of Farmoor, a few miles west of Oxford. Back at their hotel at Oxford, across the road from our sub-department of Theoretical Physics, I told him that I had achieved one of my ambitions, which was to obtain permission to place a memorial stone for P.A.M. Dirac in Westminster Abbey, the place where the British people had traditionally commemorated their great men and women, and that this stone would be put in place on November 13th. Chandra found this quite strange, and, as we ate and drank our afternoon tea, he asked me: "Dick, tell me, why do you want to do this? Tell me now," not once but a number of times. I gave him various answers but none satisfied him. I can still hear him asking this question; it is always the first memory which comes into my mind when his name is mentioned. I wish that I had stayed with them longer that afternoon.

At all stages of his life, Chandra carried a self-imposed burden much greater than any one of us would be able to bear, so much so that one felt almost ashamed to contemplate this fact. He did it without any complaint — he was always courteous, and always fair but firm. He was an inspiration to us all.

Acknowledgements

I would like to thank M. Rees, W. Israel, G. Gibbons, M. Perry, J. Barbour, P. Townsend, D. Page, and others for recollections of Chandrasekhar's opening talk at the Newton Tercentenary Conference in 1987, and especially W. Israel for letters concerning the origins and contents of that conference and D.N. Page for notes he took of Chandra's talk, apparently the only contemporary record of it, even its title. I am also indebted to K. Srinivasa Rao for details of some Indian events concerning Chandra.

Notes

1. Even if a talk were printed there, this would not settle the matter. Steven Weinberg's article in the conference book was entitled "Newtonianism and Today's Physics," a talk he explained, which he had given in April 1986 at the University of Texas at Dallas. The talk he gave at the conference was entitled "Newtonianism, Reductionism and the Art of Congressional Testimony," as we know from its publication in *Nature* [**330** (1987) 433–437].

Chandra the Romantic

Peter G.O. Freund

It is hard to start writing about Chandra in the past tense, knowing full well that he will never again show up at my office door. For many years we used to meet once a month or so, for an hour or more, whether in his office or mine. We discussed physics a lot, but our interaction covered more mundane topics as well and there was a very natural flow to all this. Physics, even in its most exalted forms, has a history and this history in turn involves personalities and, whether one knows them firsthand or not, one has a tendency of relating to these personalities in a biased manner. Now, Chandra and I did not always have the same biases and this tended to heat up the discussion and to steer it in personal directions which ultimately led us to reveal something about ourselves, as biases always tend to do. Over the years we got to know each other quite well and what developed, I came to think of as friendship. Beyond the cool, elegant scholar and lecturer, whose memory we all cherish, there was an intense, passionate, indeed romantic Chandra. It is this romantic Chandra whose creations we and future generations shall always admire, and it is this Chandra whom I discovered over the years.

Chandra, when I first met him, was just moving on to a new field, general relativity; he did this deliberately about every 10 years. Gregor Wentzel, then at Chicago, had suggested general relativity to Chandra, but the fact that Eddington had been active in this field may also have had something to do with Chandra's decision. The hero of general relativity is of course Albert Einstein and not surprisingly Einstein became Chandra's hero as well. At this point Einstein could do no wrong, as far as Chandra was concerned, and no one could fault him for this hero-worship, certainly not I. Chandra had met Einstein in Princeton. He told me that he and Martin Schwarzschild were permitted to sit in the back of the lecture hall in which Kurt Gödel presented to Einstein his troublesome solution to Einstein's equations. The presentation was in German, Einstein addressing Gödel as "Herr Gödel", who in turn addressed Einstein as "Herr Professor". The two men knew each other well, but Herr Professor Einstein was Herr Gödel's senior by more than a quarter of a century, and this is how Teutonic etiquette has it, even in New Jersey. Einstein was apparently not worried by Gödel's results, he attributed them to Gödel's use of a — in Einstein's eyes discredited — cosmological term, little did he know. What Chandra loved so much in Einstein, like most people I guess, was that unique conceptual clarity that so pervades his work. Like all other stages of Chandra's career, this general-relativistic stage led to a book, this one on black holes. By then Chandra started on his next stage devoted to Newton's *Principia*. With Sir Isaac Newton as his new hero, again a perfect choice, Chandra very suddenly turned on Einstein. Overnight, as it were, Einstein became the object of Chandra's vehement criticism. He was now faulting Einstein for not being Newton's equal as a mathematician. How could someone write down the equations of the relativistic theory of gravity, Chandra asked, and then use approximations instead of working out the exact solutions? Had Newton written down these equations, *he* would certainly *not* have needed (Karl) Schwarzschild, Kerr, Reissner and Nordstrom to find the solutions, he would have found them right away himself. All of a sudden I found myself cast as Einstein's "defender". I say "cast" because Einstein does not need me, or anyone else to defend him, the old boy does quite well on his own, thank you. Why did Chandra turn like this on Einstein, I often asked myself. The best I could come up with was that hero-worship has its dangers. When a new hero appears on the scene, his predecessor must fall. So be it. But then the beauty of all this is that, whether fair or not, Chandra's attitude toward Einstein and then Newton drove him to remarkable insights of his very own and in the end this is what counts.

Chandra became interested in Newton's *Principia* around the tercentenary of this seminal work of Western science, when he was asked to lecture and write about it. He decided to limit himself to a set of propositions which form the core of the *Principia*. He read in Newton the precise formulation of these propositions, then sat down and proved each of them for himself and finally compared his own proofs with those of the master. Knowing that Newton had invented the calculus, Chandra allowed himself free use of this mathematical power-tool. Much to Chandra's astonishment, virtually none of his proofs matched the originals. Invariably Newton came up with some awesome and totally unexpected trick. Where Chandra used standard calculus, Newton drew a "clever line" or made some other very insightful elementary geometric observation to get there faster and more elegantly. Why did Newton do this? Chandra's answer: that's how true genius does it, no approximations here ... To my mind, two alternative explanations appeared just as plausible, but these did not please Chandra. Could it be that Newton used the calculus sparingly in order to be more readily understood by his contemporaries? On the other hand, even Newton grew up before his own invention of the calculus and was very well versed in pre-calculus geometrical methods. Could it be that like most humans even Newton thought in terms of familiar old methods and only used the *new* calculus when these older methods failed him? I do not know which, if any, of all these explanations is correct, but I do know that Chandra was adamant about his view and that we argued over it for many hours.

We also often got into politics. One particular instance I remember well. It started in 1972 during the presidential election campaign. Chandra was wearing a large McGovern button on his lapel; I on the other hand favored Nixon, largely because I found the "opening to China" a bold, imaginative and long overdue American foreign policy initiative, whereas McGovern came through to me as a self-righteous individual without new ideas. Chandra saw no merit whatsoever either to Nixon's foreign or to his internal policies and was nothing short of appalled by Nixon's "tilt" toward Pakistan. Sometime that summer Chandra came to my office, McGovern button still at the ready, and told me, "Peter, I was in Germany and talked to Heisenberg. You will be pleased to know that on Nixon he holds views very similar to yours." This was the ultimate put-down of which Chandra was capable, for not only did Chandra disapprove of Heisenberg's wartime behavior, he also knew that on this count we agreed. This put-down was delivered with that familiar twinkle in his eyes, but at the time it had virtually no effect on my views. The story though does not end here. Before long the famous piece by Oriana Falacci on Kissinger became a *cause célèbre.* In it Kissinger attributed his own, then considerable, popularity to the Americans' fascination with the image of the lonely cowboy riding into the sunset. This was then generally interpreted as an attempt by the National Security Advisor to claim sole credit for all major successes in American foreign policy. Though he probably deserved such credit, it would have been more appropriate to let history give it to him. In any case, Nixon's own sizeable ego was bound to react to such a statement and I started worrying that he would take some "wild" action to "prove" that it was *he* and not Kissinger who ended the war in Vietnam. Alas, something very much like the action I feared, did occur at Christmas that year, when Nixon ordered the bombing of Hanoi. This time *I* was outraged. A few days later I ran into Chandra in the Fermi Institute library and told him so. The twinkle returned to his eyes as he replied, "If you mean what you say, why not do something about it? As it happens, I have already drafted a letter to the US Senate requesting that it censure the President's action. Want to join me?" We hurried to Chandra's office, I read the letter, signed it and his secretary then mailed it. The rest is history; the Senate, just to spite Chandra and me, did not censure anybody. But then the senators had Watergate up their sleeves, and by the time those momentous hearings started, I could watch them on television with a clear conscience, well maybe not as clear a conscience as Chandra's ...

Politicians do not hold a monopoly on ego. Ego can interfere — constructively, or destructively — with the work of physicists as well. This theme of ego in physics arose often in our conversations. I vividly recall Chandra's story about the Oxford cosmologist E.A. Milne. Chandra had once happened upon Milne standing in deep thought in front of a blackboard on which, written below each other, were four names: Newton, Maxwell, Einstein, Milne! I can still hear Chandra asking, "Can you imagine this? Isn't it sad?"

Maybe the most remarkable thing about Chandra was the total honesty with which he was able to assess his own strengths and weaknesses. It is this ability that always put him on the optimal path with a great deal of certainty and commitment. When he retired, I asked him whether he considered returning to India. No, it would make no sense, he replied without any hesitation. Sensing that his certainty had surprised me, he brought up the name of D.S. Kothari, his distinguished Cambridge classmate, who decided to return to India, where he played a leading role in building a first rate school of physics. Had Kothari stayed abroad, who knows, maybe he would have achieved even more as a scientist, but he knew he had to go back. Chandra, on the other hand, felt that he himself was neither interested in nor particularly skilled at organization and administration and therefore his only meaningful alternative was to stay in Chicago and continue doing what he did best, research that is. Then he added philosophically, "In the long run, maybe Kothari's contribution will turn out to have been the more important one by far."

S. Chandrasekhar, the Friend: Some Reminiscences

Agnes M. Herzberg

"It is hard to imagine a world and life without Chandra." This is how my father began his letter of condolence to Mrs. Chandrasekhar. He went on: "For the rest of my life I shall remember the frequent joint walks that we had during my time at Yerkes and the many discussions we had on scientific and many other topics. Needless to say I learned immeasurably from him." This sums up the effect Dr. Chandrasekhar has had on many, including my family and me. His influence was deep.

Subrahmanyan Chandrasekhar was the good friend of my father, Gerhard Herzberg. I first met Dr. Chandra, as we children called him, and I continued to do so, in August, 1945, when my family moved to Williams Bay, Wisconsin, where Yerkes Observatory of the University of Chicago is situated. I had, therefore, known him for fifty years. Dr. Chandra was our neighbor and my father's colleague. At the time, I had no idea of his stature, even then, in the scientific community although I did notice that my parents seemed to want to get everything perfect before Dr. and Mrs. Chandra came to visit for the evening or when they came to stay with us later, after we had moved to Ottawa. To me, he was a friend and playmate.

During our time in Williams Bay, he told me many stories, including how to remember his birth-date, how to add easily the numbers from zero to one hundred and the Principle of Maximum Unhappiness. One of my earliest memories is of him, dressed in black on Halloween, chasing us "observatory brats," a term coined by my mother for all the astronomers' children. Further, on a day-long picnic on a very hot day, Dr. Chandra was the only one whose shirt-sleeves were uncreased when rolled down. This, in our family, made him seem like "magic," and the term stuck; often magic seemed best to describe what happened around him.

In 1948, when my family left Williams Bay, Dr. Chandra wrote in my autograph book "For Agnes, who will come to Yerkes to study astronomy in 1958." I did not go to Yerkes to study astronomy, but I went to Queen's University to study mathematics instead. I fear I would not have been able to stay the course.

In spite of my not going to Yerkes Observatory, Dr. Chandra became a friend and mentor; the person I went to many times for advice on all sorts of problems, both scientific and personal. He was like an uncle to me and a true friend, an advisor and a confidant. He always made time and was there in one's hour of need.

Winston Churchill advised the young "to know how to command the moment to remain" (Churchill, 1967, p. 98). Dr. Chandrasekhar was a past master of this; unlike ordinary mortals, his memory was as perfect as he was! One could question the authenticity of what he said, but my father has said in his introduction to a lecture given by Professor Chandrasekhar in Ottawa in November, 1989:

> Chandrasekhar is a man of extraordinary wide interests and an extraordinary memory. Much of the detailed history of physics, astronomy and mathematics is embedded in his memory. I learned from him for example that Eddington [Sir Arthur Eddington] said, "You cannot believe in astronomical observations before they are confirmed by theory." He told me the story of Eddington, a Quaker, and conscientious objector in the First World War, who took a leading part in the eclipse expedition in May 1919 which gave the first direct confirmation of the predictions of general relativity. He told me the story of Raman [C.V. Raman] getting the Nobel Prize (never mentioning that he was Dr. Chandra's uncle) and when years later, I met Raman in Bangalore he told me the same story in almost the same words.

Dr. Chandrasekhar used anecdotes to explain and illustrate and, in so doing, told one what he thought. In November, 1989, the National Research Council of Canada held a meeting in honor of my father's 85th birthday. My father worried that Dr. Chandra would not be looked after to both my father's and Chandrasekhar's satisfaction. My father, therefore, made me Dr. Chandra's personal chauffeur. I picked him up at the airport; as we drove into Ottawa, he told me the following story.

Dr. Chandra had organized a meeting at which Henry Norris Russell was present. Russell, it seems, drank a lot of water during every meal and

Dr. Chandrasekhar was up and down all the time keeping the pitcher full and doing other errands. Finally Russell said: "You need not treat me like a maharajah." I believe Dr. Chandrasekhar was telling me that he did not feel that he was so special as to receive special treatment.

There are many stories in Dr. Wali's book that always remind me of the stories Dr. Chandra told me. Chandrasekhar made his students work hard and when students could not see the point of continuing would tell them the following (Wali, 1991, p. 18).

> There were five princes. When they were taking archery lessons from a famous master, one of the five princes became known as the greatest of them all. On one occasion, a visitor — a wandering minstrel — comes to the archery school and sees the five princes practicing. To him all of them appear extraordinarily good, nothing discriminates one from the other. When he encounters the master with his observation and asks him why one is picked as the greatest, the master leads him to the five princes. The master asks each prince to take aim at, but not shoot, the eye of a bird sitting on a tree. When they are ready, he asks each of them, "What do you see?" The first prince says he sees the bird's eye, the tree branches, flowers, and the sky beyond. The second prince narrows the list somewhat, but when it is the turn of the prince who is known to be the best archer of them all, he says, "Revered Master, it's strange. I don't see anything except the eye of the bird."

One should have supreme concentration for the task at hand.

Another point that Dr. Chandrasekhar reiterated many times was that only posterity can decide how well one has done. One's peers cannot do that even if they think they can. He had learned this from E.A. Milne. Wali, in his book (1991, p. 297), quotes Chandrasekhar on Milne:

> ...posterity, in time, will give us all our true measure and assign to each of us our due and humble place; and in the end it is the judgment of posterity that really matters. ...He really succeeds who perseveres according to his lights unaffected by fortune, good or bad. And it is well to remember that there is in general no correlation between the judgment of posterity and the judgment of contemporaries.

When Dr. Chandrasekhar received an honorary degree of Doctor of Science from Queen's University in Kingston, Canada, in June, 1991, I wrote the citation given in Figure 1.

Figure 1. Citation for Honorary Degree from Queen's University

At present Distinguished Service Emeritus Professor at the University of Chicago;

Where he has professed his knowledge for over fifty years and by coincidence or chance whose university celebrates its centennial year;

Indian by birth, American by adoption, who like Columbus made the voyage to the New World;

Graduate of Presidency College in Madras and of Cambridge University, and whose scholarly achievements are attested by Fellowship in the Royal Society of London at a very early age and culminating in the Nobel Prize;

A leader in pushing the frontiers of science in many directions including the mathematical modeling of outer space, thus inciting physicists, astronomers, mathematicians and statisticians to claim him as their own;

Prodigious writer of scientific papers and monographs who also knows the beauty of Shakespeare, Beethoven and others;

Whose sense of humor and extraordinary memory of scientific anecdotes provide antidotes for all:

Teacher and lecturer about whom it has been said, "He is a grand master who with elegance, grace and scholarship, literally charms his audience and keeps them spellbound;"

And whose famous class consisting of just two students, who together won the Nobel Prize, questions the concept of cost-effectiveness in universities;

Discriminating editor who with wide knowledge and sure acumen guided *The Astrophysical Journal* for nineteen years;

Known also for his eponymous limit which exists for the stars, white dwarfs and black holes;

Whose own excellence and devotion to the world of science and humanity surely sets an upper bound to what is possible;

And today, by honoring one of the twentieth century's most eminent scientists and a living legend, Queen's itself has been honored with a jewel in its sesquicentennial crown.

Dr. Chandra told me he liked the citation. The day before the ceremony he gave a lecture and on that evening there was a large party at which he and my father participated in a little impromptu skit. The picture is from that party.

My father wrote the following on the occasion of the 65th birthday celebration of Dr. Chandrasekhar; it describes their relationship very well.

> I first met Chandra at an informal meeting at Yerkes Observatory back in 1941. At that time Chandra was rather more reticent than he is today and I did not have much contact with him. I do, however, remember one evening after the meeting when Chandra and Lalitha helped me find in the complete darkness the way back to my cottage in the YMCA camp [George Williams College Camp].
>
> My real acquaintance and friendship with Chandra started in 1945 when I joined the faculty at Yerkes Observatory and was assigned an office directly opposite to Chandra's office. To me the close association with Chandra was the most rewarding result (scientifically and personally) of the years I spent at Yerkes Observatory.
>
> During this time I learned what a prodigious worker Chandra was, how clearly his aim was always in front of him and how immense his influence was on his students (as it was on me) both by his example and by his teaching (in lectures as well as discussion).
>
> Sometimes when Chandra felt he needed a break he came over to my office and we had a long chat either in my office or on a walk along the shore of Lake Geneva. We usually discussed some phase of his or my work or indulged in some reflection of the history of our subjects. Chandra with his incredible memory was able to recite almost verbatim some of the conversations he had with Eddington, Milne, Fowler, Rutherford, Chadwick, Blackett, etc. during his Cambridge years which coincided with one of the most exciting periods in physics. It is good to know that some of these historical items have been preserved in Chandra's more general lectures. ...
>
> What we admire in Chandra is not only his intellectual power, the scientific results of his work, but equally his humanity, his sense of fair play, his concern for his fellow men, his undaunted spirit and his warm friendship. He has in an exceptional degree a sense of perspective which is combined with a fine sense of humor.
>
> As an example of his sense of perspective I like to refer to his convocation address at the University of Chicago in 1971, in which he tried to answer the question of a student, "How can you justify spending all your efforts on abstract astrophysical problems when there are untold millions in India in dire poverty?" There is no question about the validity of his answer: Survival is certainly a necessary aim of society, but

> without at the same time keeping alight the flame of the highest intellectual endeavors it (i.e. survival) is not worth the effort. Chandra did not express it quite as bluntly as I just did but far more eloquently and almost poetically. Whichever way you like to formulate this answer it is our duty in our egalitarian society to emphasize this, to us, obvious truth.

The story of Dr. Chandra always told to me concerning my mother was that during our time in Williams Bay, my mother, also a physicist in her own right, went to a course of lectures given by Chandrasekhar. Dr. Chandra told her that if she came to every lecture, he would give her a prize. Being very modest and shy, my mother did not go to the last lecture in order not to be singled out.

Recently I have seen the following quotation by K.F. Gauss:

> It is not knowledge, but the act of learning, not possession but the act of getting there, which grants the greatest enjoyment. When I have clarified and exhausted a subject, then I turn away from it, in order to go into darkness again; the never-satisfied man is so strange — if he has completed a structure, then it is not in order to dwell in it peacefully, but in order to begin another. I imagine the world conqueror must feel thus, who, after one kingdom is scarcely conquered, stretches out his arms for others.

Gauss and Chandrasekhar had the same style of work.

My father ended his letter to Mrs. Chandrasekhar with: "It is our good luck that Chandra was able to complete the translation of Isaac Newton's great work which will be immeasurably useful to generations of those interested in the history of physics."

In November, 1994, I stayed with the Chandrasekhars for a few days; a cousin of mine, Marion Thurnauer, was there also for the first lunch. After some introductory conversation about Newton, my cousin said: "I understand that Newton was not a nice man." Dr. Chandra promptly, rising from his chair, said: "What do you mean, Newton was not a nice man?" We had to spend some time on the meaning of "nice" and "not nice" and how one can say that about individuals.

When I returned to Canada, I telephoned my cousin. She said that when she left after lunch, Dr. Chandra had not only taken her to the door of the apartment building, but also to her car. "Well," I said, "he is a nice man." She replied at once: "What do you mean by 'nice'?"

I have recently reprimanded my father for not having told Paul, my brother, and me when we were children to always have our notebooks out when Chandra visited, but I fortunately remember much.

J. Cronin, Dr. Chandrasekhar's friend and neighbor, was completely correct when he said: "He is the closest form — closest Platonic form — of an ideal individual; a truly great scientist, one among a very few in this century" (Wali, 1991, p. 33).

I was fortunate indeed to have Dr. Chandra as a friend; it is my feeling that I was blessed with a very special relationship. For me personally and for the world in general, he cannot be replaced.

Notes

1. Churchill, S., *A Thread in the Tapestry* (London: Andre Deutsch, 1967).
2. Wali, K.C., *Chandra: A Biography of S. Chandrasekhar* (University of Chicago Press, 1991).

A Towering Figure: Reminiscences and Reflections

L. Mestel

I think I heard Chandra's name for the first time in 1948, when as a third-year Cambridge undergraduate I attended Fred Hoyle's lectures on "Statistical Thermodynamics", in which he introduced us to the Fermi–Dirac statistics and ultimately to degeneracy, non-relativistic and relativistic. At about the same time I recall Fred's writing a review of the posthumously published "Fundamental Theory", in which Sir Arthur Eddington referred to the "havoc" being wrought in astrophysics by the Stoner–Anderson–Chandrasekhar equation of state. Fred had no option but to point out that Eddington was virtually alone among theoretical physicists in his opposition (expressed I believe even more forcefully at the 1939 Paris meeting on white dwarfs etc.). Already, Fred's pioneering work in nucleogenesis had shown that, unlike the non-relativistic Fowler equation, which Eddington wanted to hold over the whole parameter range, the relativistic equation of state ensures that evolving massive stars contract to states of such high density and temperature (the "hotter place", demanded by Eddington himself in a famous riposte) that high mass elements can be synthesized. And of course once I began working as Fred's student on stellar structure and evolution, I was continually dipping into both Eddington's "Internal Constitution of the Stars" and Chandra's "Introduction to the Study of Stellar Structure". And like Fred and others, I had no difficulty in feeling admiration for both Eddington and Chandra, while remaining puzzled at what seemed at the least Eddington's violent overstatement of his case.

This ongoing controversy between Chandra and Eddington not only put a strain on their relationship while Eddington was alive, but undoubtedly coloured Chandra's memory for decades afterwards. Half of him could not forget what he considered Eddington's devious behaviour at the time of the now famous January 1935 meeting of the Royal Astronomical Society. Dennis

Sciama tells me that as late as 1972 when he was on sabbatical leave in Oxford, Chandra spent a whole post-prandial dessert at All Souls' College recounting his version. To be fair, one should note that others — such as Sir William McCrea — put a very different interpretation on Eddington's motivation. But since this image of persistent resentment is very widespread, it is pleasant to be have personal evidence of Chandra's later reconciliation with Eddington's memory.

Let me digress. In the "Internal Constitution", Eddington (admittedly with some sleight of hand) derived his relation predicting that the luminosity L of a homogeneous star should depend strongly on its mass M but only weakly on its radius R. A newly formed star would contract, with gravitational energy supplying the Eddington luminosity and simultaneously increasing the internal temperature like M/R, until nuclear processes begin to release energy. The stellar "Main Sequence" is defined by stars of different mass, each with the internal temperatures at the right level for nuclear energy generation to balance the luminosity. Eddington noted that the observational material to hand not only confirmed a strong L–M dependence, but also yielded only a modest increase with M of the central temperature: he therefore inferred a strong temperature sensitivity for the nuclear energy release, a prediction vindicated years later. During a visit to Yerkes in 1955, I remarked to Chandra what a remarkable coup this was, the stuff of personal Walter Mittyish type dreams. There had been no disagreement between Chandra and Eddington on the theory of non-degenerate stars. I was therefore unprepared for Chandra's dismissive response: "Oh it was only M/R." I could not help wondering whether Chandra's lingering resentment over the degeneracy controversy had spilled over.

Twenty-five years elapsed. Over lunch with Chandra in Chicago, I remarked that I had listened to a recent broadcast in which John Maddox (editor of *Nature*) interviewed him, in particular reviving the controversy of the 30's. Again to my surprise, Chandra broke in: "You people have got it all wrong: I had a very good relationship with Eddington. Look what a remarkable man he was: consider his prediction from his theory that nuclear energy liberation must be strongly temperature-sensitive ...!" On my reminding him of our earlier interchange, he said: "Well, I've grown wiser with the years." There was certainly no lack of generous recognition in his centenary lectures on "Eddington: The Most Distinguished Astrophysicist of His Time".

Chandra's style of working was very personal. He would move into a new area, master the basic physics and its mathematical formulation, and then proceed to make important — often outstanding — contributions to its mathematical development. In his extensive work on radiative transfer, Chandra's ingenuity in dealing with the resulting integral equations has excited the admiration of pure mathematicians such as my Cambridge colleague Frank Smithies, who abstracted many of Chandra's early publications in the field for *Mathematical Reviews.* The series of monographs tell the story of his scientific life: *Stellar Structure*; *Principles of Stellar Dynamics*; *Radiative Transfer*; *Hydrodynamic and Hydromagnetic Stability*; *Ellipsoidal Figures of Equilibrium*; *The Mathematical Theory of Black Holes*; and *Newton's Principia for the Common Reader.*

Chandra chose to work on topics which originated through the development of astronomy over the centuries. After the trauma of the controversy with Eddington, he must have been gratified when on the one hand the discovery of neutron stars, and on the other the further development of Fred Hoyle's work in nucleogenesis, confirmed that relativistic degeneracy must be an essential part of our picture of stellar evolution. Yet it is clear that it was the mathematics that became for him the overriding challenge: the attempt to understand in detail what we see in the heavens he just preferred to leave to others. And if over the years an area of study had become remote from its original astronomical motivation, this was of far less importance to Chandra than the opportunity to exploit a mathematical gold-mine. This is shown probably most clearly in the work on rotating, self-gravitating liquid ellipsoids, with or without internal motions. Problems which had attracted the efforts of so many eminent mathematicians, starting with Newton and McClaurin and including Jacobi, Dedekind, Riemann and Poincaré, look now like a tailor-made challenge to Chandra and his collaborator Norman Lebovitz. It is with characteristic insight that Chandra saw that the elaborate formalism developed by previous workers had made the subject unnecessarily complicated, and that the most important results could be obtained in Cartesian coordinates. It is a delight to witness the ease with which Chandra and Norman populate the parameter space, in particular deriving points of bifurcation of one series of equilibria from another. The link with astronomy, though by now tenuous, is not absent: Ken Freeman's galactic studies had produced analogues of the Riemann ellipsoids, with the vorticity vector of the internal motions of sign opposite to the rotation vector of the ellipsoidal figure. But Chandra I am sure looked on this just as a mildly gratifying spin-off, certainly remote from the motivation that led to the whole monumental exercise.

Examples come to mind which show how Chandra was sometimes ahead of his time: his mathematical expertise led him to results which were not appreciated by the astronomical community till later. Once the basis of magnetohydrodynamics had been established, it was inevitable that Chandra would explore some of the consequences in mathematical detail. He was the first to write down the most general exact, steady-state integrals for a rotating system with a magnetic field symmetric about the rotation axis. It was left to others in fact to re-derive the integrals and show how they could describe the rotational history of the Sun and other solar-type stars, with possible spin-off regarding the origin of planetary systems. In his volume on hydrodynamic stability, there is a result that confirms one's intuition that the presence of a magnetic field could alter the whole stability problem for a Keplerian disc of gas. This is now the starting point of ongoing studies of magneto-turbulent "accretion discs" surrounding pre-main sequence stars and neutron stars.

Again, I think that one primary reason for his original move into general relativity was his conviction that people had been mesmerized by Einstein's non-linear equations. His development of the "post-Newtonian approximation" scheme was in fact opportune, for the discovery of the binary pulsar had opened a laboratory for studying the implications of general relativity, even if still in the "weak-field" domain that allows an expansion procedure. Recall that one of the classical tests of general relativity is its clear prediction of the 43 seconds of arc per century perihelion advance observed for the planet Mercury. The discovery of a system for which the analogous effect is 4 degrees per year and for which other characteristic g-r effects are likewise scaled up left the community of relativists somewhat dazed. In particular, Joe Taylor's continual monitoring of this system confirms to astonishing accuracy the spin-up of the orbital motion through energy loss as gravitational radiation. But for Chandra it must have been natural to move on from his studies in depth of Newtonian gravitation to face the challenge of g-r. He reproved the applied mathematical community for its lack of courage: in his own words, there was no reason why the post-Newtonian expansion could not have been developed years earlier. This work, along with that by Hermann Bondi, Andrej Trautman, Kip Thorne and others, laid to rest the lingering doubts about the reality of gravitational radiation, and presumably has played at least a minor role in spurring on the rival groups building detectors that will hopefully absorb such radiation emitted by a variety of cosmical objects. But who other than Chandra would have then gone on to produce the work for his penultimate massive monograph on black holes; including labours of love like the solution of Dirac's equation to the electron in the Roy Kerr's metric for a rotating black hole?

Contemplation of Chandra's achievements leaves one filled with admiration (not a little tinged with envy). But the question does inevitably present itself: had I been blessed with his mathematical gifts, how would I have used them? Is it really a piece of chutzpah to say: I personally would have got more satisfaction — and indeed would also have contributed even more to the scientific community — by pausing every so often to devote just a fraction of my time to the direct advancement of astronomy. I am not surprised that others, such as my late friend and mentor Tom Cowling and my former student and colleague Donald Lynden-Bell, should have expressed similar thoughts. And I believe that Chandra himself did look back and ask whether his scientific priorities had always been the best. He could not but note that the late Jan Oort, with far simpler mathematical tools than those used by Chandra in his work on stellar dynamics, but with his eyes firmly on the observational material, could introduce the Oort constants describing the rotation of our galaxy — household words for the working astronomer — and could start the ongoing hunt for "missing matter".

One lesson is that every gift has its potential dangers. It has been said that in 1915 there were many people, e.g. in Goettingen, who were technically better mathematicians than Einstein, in particular knowing far more about non-Euclidean geometry; but to produce the new conceptual framework leading to g-r, one needs to think of mathematics as a servant rather than a master, and to be guided by the physics. This is not meant to imply for a moment that Chandra had less physical insight than most of his contemporaries of comparable stature, but rather that his instinct was to trust his mathematics more than his physical intuition.

Of course a man has the right to spend his one life as he wishes. And perhaps the answer to Chandra's occasional self-doubts has already been given by Alexander Pope: "Man never is, but always to be blest."

Encounters with Chandra

Jayant V. Narlikar

It was a warm summer morning in July 1962, with time coordinate around 7.30. The space coordinate: a country house (which was used as a conference centre) in Jablonna, near Warsaw. The occasion: the Third International Conference on General Relativity and Gravitation. I was attending the meeting as a research student working in cosmology under Fred Hoyle at Cambridge.

It was too early for breakfast and so I was taking a stroll through the spacious and well-kept grounds of the estate. There I saw another Indian similarly occupied. He was in his fifties, distinguished-looking, and dressed immaculately in a dark grey suit. We greeted each other and introduced ourselves.

That was my first meeting with Professor S. Chandrasekhar, or *Chandra*, as he became popularly known. But what was he doing at a relativity conference? I knew of him as *the* Chandrasekhar of the celebrated *Chandrasekhar limit* of the white dwarf stars. I had also known of his work on stellar dynamics. Perhaps he saw my puzzlement on my face, for he went on to explain that he was attending the conference to find out what were the interesting areas to pursue in general relativity, a field that he wished to enter anew.

I could not but admire the decision, which I termed to myself as *adventurous*, for was it not so to embark on an entirely new research area in one's fifties?

During the conference Chandra did not present a paper but listened quietly to the stalwarts holding forth on gravitational radiation, exact solutions, cosmological models, geometrical classification of spacetimes, etc. etc. There was also some discussion of spinning models, a topic that had been introduced into relativistic cosmology by the classic paper of Kurt Gödel in 1949 in the *Einstein Seventieth Birthday Volume* of *Reviews of Modern Physics*. From Chandra's subsequent work this paper seems to have excited him ... as I discovered the following year.

* * *

In the spring of 1963, there was an unusual meeting hosted by Tommy Gold at the Physics Department of Cornell University, which I had the opportunity to attend. I was still a research student and as luck would have it I was to appear for my viva voce for Ph.D. at this meeting!

The meeting was on "The Nature of Time" and was conducted round a long table amongst a few selected invitees. These included such luminaries as John Wheeler, Dick Feynman, Fred Hoyle, Hermann Bondi, Dennis Sciama, Phillip Morrison, Adolph Grunbaum, etc. And, of course, Chandra, who now had something to report at the meeting. His contribution, amidst a quasi-scientific and quasi-philosophical discussion, was characteristically *exact.* He had worked out the geodesics in Gödel's universe and had found that the much-publicised claim of existence of closed timelike lines was not correct in the sense that these trajectories did not allow one to travel into one's past.

It was at this meeting that he invited me to visit Williams Bay whenever I was passing through the Chicago area.

* * *

Opportunity to visit Chandra came not on that trip but six months later when I was again in the United States to attend the Dallas Symposium (the first of what subsequently became known as the Texas Symposia) on the newly discovered quasi-stellar objects, the phenomenon of gravitational collapse and the whole new area of relativistic astrophysics that had opened out. I wrote to him enquiring the possibility of visiting him and he readily invited me to give an account of the conference in the form of a seminar.

Thus in early January I took a train from the Chicago station for Williams Bay, following the meticulous instructions sent by Chandra. Chandra was at the other end to meet me and we went straight to lunch at a pleasant restaurant. On the way Chandra casually mentioned to me that at seminars he became quite critical of the work presented, but not to mind it as it was not personal. He mentioned an anecdote: "When Professor X was here giving a seminar on his work I kept on interrupting him with queries as to how he got there, because I was not satisfied with his reasoning. As he was feeling increasingly cornered, he finally burst out saying that an elephant may trample a fly and say that he did not know how the fly got there."

It was a sobering thought! I hardly had much experience of giving seminars; but one comforting thought was that here I was going to describe not my own work but that presented by others at the symposium.

During the seminar there was one occasion when Chandra became quite aggressive. I was describing Willy Fowler's argument using the gravitational binding energy curve of a supermassive object to show how it leads to the onset of gravitational collapse. Chandra's comment was that such arguments were imprecise: that one must do a small-oscillations analysis in order to decide if an equilibrium configuration was stable or not.

Sure enough, within a few weeks a paper by Chandra appeared in *Phys. Rev. Lett.* carrying out the small-oscillations analysis and demonstrating instability *precisely* and *rigorously*.

But, to return to that day in Williams Bay. After the seminar Chandra took me to show the place of his other incarnation: the editor's office for *The Astrophysical Journal.* This was the place from where the journal was being shaped, almost single-handedly by Chandra, from a journal of moderate reputation to one that brought it to the top section of the league of science journals. He described how he divided his time between his own teaching and research and the management of the *Ap. J.*

Then Lalitha joined us and we went to their home for a South Indian meal, one I was having after a long time. Here Chandra's reserve melted and we had a nice chat. I was, however, anxiously looking at my watch for the train back to Chicago.

"Don't worry. I have the timetable worked out: I will get you onto the train on time," assured Chandra. Faith though I had in his mathematical precision, having been brought up in an Indian environment where there are many imponderables, I kept worrying secretly. But here, like the equations and boundary conditions that Chandra concerned himself with, everything was under control and he brought me to the station a couple of minutes before the train rolled in.

* * *

I never met Chandra in Cambridge. Even when I returned to India I had very few encounters with him. But those few that I had were worth remembering.

Like in the early 1980s when he visited the Tata Institute of Fundamental Research at Bombay. I invited Lalitha and him for tea at our flat on the campus. My parents were staying with me and my father was meeting Chandra after five decades. They were contemporaries at Cambridge in the early 1930s when my father was a student of Eddington.

I remember that reunion as a delightful occasion with some exchanges about old Cambridge personalities. It seemed as if the gap of fifty years had never existed at all.

Later that evening there was a dinner hosted by TIFR at which I was sitting next to Chandra. He gradually opened out in his conversational topics sharing his experiences about managing the *Ap. J.*, about his policy of selecting his research areas, about students, about his reasons for staying abroad after completing his doctorate at Cambridge ... I could catch glimpses of the human being behind the formal and reserved demeanour.

* * *

During his visits to India there were two occasions when I had the chance of interviewing Chandra on TV: in 1984 at TIFR and in 1992 at Pune. Of the many topics that we talked about in 1984, was one on black holes about which he had just spoken in his Birla Award Lecture in the most glowing terms. While the decade just over had seen the black holes rule high energy astrophysics with their being quoted as panacea for numerous astronomical observations, what made them "beautiful" in Chandra's eyes was their mathematical symmetry. Indeed he had just brought out his book *The Mathematical Theory of Black Holes*. Where do you go from here ? I asked. Now that he had closed the books on black holes which new area was he going to tackle?

"Well, I am in my seventy-fifth year, which is hardly the stage for someone to take up a new subject ... but ..."

And he went on to describe his ideas on colliding gravitational wavefronts! It was a subject daunting enough for someone fifty years younger; but he did continue working on it. And in 1992 when at IUCAA (Inter-University Centre for Astronomy and Astrophysics) I invited him to deliver a lecture at the dedication ceremony of the Centre's buildings, he chose a topic which drew heavily on this work. He compared the beauty of Claude Monet's Series Paintings with the symmetries and similarities of special solutions in general relativity that described wildly different objects like the spinning black hole and colliding gravitational waves!

But 1992 already saw him embark on yet another venture. While he was extolling the beauty of general relativity and the genius of Albert Einstein, he was engaged in the task of explaining the grandeur of another scientific work. In 1992 he spoke at IUCAA also on: *Newton's Principia and Its Relevance for a Student of Today.* In private conversation Chandra regarded Newton as a greater genius than Einstein. In the lecture Chandra's adulation of Newton came through clearly, on how the great man had proved the propositions in the *Principia* so elegantly using the now outmoded methods of geometry, on how the Newtonian solutions were superior to Chandra's own alternative proofs using techniques of calculus (which Newton himself had invented but forbore to use so as to make his work more understandable to his contemporaries). And, of course, all this was to form the part of his last book.

But age was catching up ... as was evident from the slowness of the body which could no longer match the ever-agile brain. There were other signs too. Like when getting up to deliver his Dedication Lecture, he fumbled and kept looking for his glasses until it was pointed out that he already had them on!

The 1992 visit was his second one to IUCAA, the first one being three years earlier when we had just started at Pune under rather primitive conditions. But even in 1989 he could see what was to come up and encouraged us on. He was visibly pleased to see all the buildings in place in 1992. More encouraging to us was the optimism he expressed in his 1992 TV interview about the potential of Indian universities and the salutary outcome of the universities sector being provided with new opportunities at places like IUCAA.

* * *

This brings me to the question of Chandra's interaction with and influence on the astrophysics community in India. The question of his settling abroad has been extensively discussed. Was he ignored by the powers that be in India in the thirties and forties when he could have been persuaded to return? Would he have been as effective in his own work had he come back?

On the first question I have heard of at least two serious offers made to him in the forties to work in India. The first was from Pandit Madan Mohan

Malaviya, founder of the Banaras Hindu University, and its Vice-Chancellor Dr S. Radhakrishnan to set up an astronomy observatory and work on the B.H.U. campus. The second came from Dr Homi Bhabha, founder of the Tata Institute of Fundamental Research, Bombay, inviting him to be a professor in the newly set-up institute.

The answer to the second question may hold the key to why he did not accept such invitations. He may have felt that within the Indian environment with its officialdom and social pressures his work might not flourish. Although I did not have any direct discussion with him on this issue his passing comments nurtured this impression.

But the budding astrophysics community in India had always looked up to him as a source of inspiration and he in turn had been sympathetically concerned with its welfare. His visits to the native land were not frequent but on those occasions encounters with him had always been worth cherishing.

Reflections on Chandra

E.N. Parker

A common interest in the large-scale effects of hydrodynamic turbulence was the basis for my first meeting with Chandra, in the Winter of 1953. I had a position as assistant professor in the Department of Physics at the University of Utah, and I had had some correspondence with Chandra on the effect of turbulence in suppressing Jean's gravitational instability. Chandra invited me to stop by the Yerkes Observatory at Williams Bay, Wisconsin on a trip to the east coast. There was a direct rail connection to Williams Bay in those days — steam-powered. I was in awe of the internationally famous Yerkes Observatory in general and of meeting Professor S. Chandrasekhar in particular. Chandra soon put me at ease and I was immensely impressed with his interest in my simple ideas and his willingness to discuss them with me. It took time away from his own research, but he seemed to feel that my interest in the subject was sufficient to justify his attention.

Evidently, Chandra was not too badly impressed because he later mentioned me when John Simpson asked Chandra who he might recommend for a position of research associate to explore the theoretical implications of Simpson's observations of the variation of the cosmic ray intensity. Simpson was the scientific leader in using the observed energy dependence of the cosmic ray variations as a probe of conditions in interplanetary space (before the space age). I was happy to accept Simpson's offer, and my wife and I came to Chicago in June of 1955.

To make a long story short. Simpson was able to show from the energy dependence of the cosmic ray fluctuations that interplanetary space is dominated by magnetic field rather than the conventional (at that time) electrostatic field to which the variations were generally attributed. The key to the problem was that the thermal ions and electrons in space are sufficient in number to constitute an electrically neutral plasma (fluid) on large scales. On this basis the idea of a solar corona extending past the orbit of Earth, proposed by Sydney Chapman, and the observed fact that space is filled with outgoing solar corpuscular radiation, pointed out by Ludwig Bierman, led to the theoretical prediction of the hydrodynamic expansion of the tightly bound solar corona to form the supersonic solar wind in 1958. It was here that Chandra entered the picture again. My paper pointing out the unavoidable hydrodynamic expansion of the solar corona to supersonic speeds at large distance, which I called the solar wind, was submitted to *The Astrophysical Journal*, of which Chandra was the editor (1952–1971). I was not greatly surprised by the adamant but nonspecific negative view expressed by the first referee. In discussing the solar wind concept with colleagues I had found only generally skeptical attitudes. When Chandra sent the paper to a second referee, he obtained the same result. It is a measure of the time and effort that Chandra devoted to his work as editor that he came to my office one day to discuss the paper. It was evident that he had read my paper and he said, "Now see here Parker, do you really want to publish this paper?" I replied that I did. He said, "I have sent the paper to two competent referees, both distinguished researchers in the field, and they both say that the paper is wrong." I replied that neither referee was specific as to what was wrong, so that I had no basis for honoring their opinions. Chandra said, "So you want to publish it." I said, "Yes." After a moment of thought Chandra said, "Alright, I will publish it." I wonder what would have been the outcome if a lesser man had been editor. I was "nothing and nobody" in those days. One must not forget that distinguished and influential expert referees can be intensely annoyed by an editor who does not follow their "infallible" recommendations. I am sure I would eventually have succeeded in getting the solar wind paper into print. But only after substantial delay and likely in a journal of less prominence.

As a jocular note it was about that time that the "phantom" research paper appeared, circulated as a reprint in the familiar format of *The Astrophysical Journal*. The proper reference to the reprint would be "S. Candlestickmaker, 1957, On the imperturbability of elevator operators. LVII *Astrophys. J.*; **237**, 476". The address of the author is given as Institute for Studied Advances, Old Cardigan, Wales, communicated by John Sykes and received by *The Astrophysical Journal* on the date October 19, 1910. The paper treats elevator operators when the occupation number of the elevator is zero, leading to the conclusion that the value of the determinant vanishes so that the matrix of the elevator operator cannot be inverted, etc.

If one consults *The Astrophysical Journal* volume and page claimed on the reprint, one is not surprised to find another paper, on fluctuation theory of polarization rather than the imperturbability of elevator operators.

The caper is reminiscent of an abstract that appeared many years ago in the APS bulletin for a meeting of the Physical Society, in which the authors proposed that the number 440 is fundamental in nature. They pointed out that (a) 440 cps is the reference frequency of the musical note A above middle C, (b) 440 v is the line voltage of secondary power lines, and (c) 440 keV is the energy of an important excited state of the 7Be nucleus. Needless to say, no one showed up at the appointed place and hour to present the paper.

As editor Chandra frequently had to cope with temperamental authors and temperamental referees. On one occasion a Nobel laureate submitted a paper for publication in *The Astrophysical Journal* with the stipulation that the paper was not to be refereed. Chandra (who was not a Nobel laureate himself at the time) replied that it was established policy to referee all papers, and that he would see to it that the refereeing was accomplished in a week so there would be no significant delay in publication. The Nobel laureate withdrew the paper. One can only guess his motives.

For some years two well-known astronomers, who occupied offices on the same corridor at the Mt. Wilson Palomar Observatory headquarters in Pasadena, were locked in a feud over the nature of galaxies. Each wrote papers critical of the other which were submitted to *The Astrophysical Journal.* Chandra often sent the paper submitted by one to the other to be refereed. The two did not speak to each other in Pasadena; their only communication was through the editorial office in Chicago.

On more than one occasion Chandra made the point to me that a referee's report contains valuable information — even the most naive and negative report. The referee is, after all, a typical reader so the referee's misunderstandings of the paper (reflected in the referee's report) indicate to the author where the exposition needs to be clarified and generally sharpened up, what points need to be explained or expanded, what additional information or calculations might strengthen the arguments, etc. Experience has borne out the importance of Chandra's point. First of all, many referees are conscientious and constructive in their comments and criticism, and in many cases help avoid unclear or erroneous statements. Second, I know of some marvelous exceptions to Chandra's statement, e.g. the referee of a paper submitted a couple of years ago by a young research associate to a journal in the field of geophysics, in which the editor turned down the paper simply because "the mathematics was not sufficiently original" even though the paper presented a whole new physical view of the geomagnetic substorm. In another case an excellent theoretical paper, providing an elegant and ingenious formal analytical solution to the magnetohydrodynamic alpha omega-dynamo equations with time varying alpha and omega (to represent the generation of magnetic field in a "starburst" galaxy), was turned down because the referee noted that the equations were amenable to numerical solution, leading to a much shorter publication, with which the editor fully concurred. That such inane remarks are made by anonymous and influential referees is not surprising, but that experienced editors should concur is astonishing, emphasizing again the importance of editors of Chandra's scientific stature and integrity.

The point of this digression is that I once had to remind Chandra of his own words of wisdom. I had submitted a paper to *The Astrophysical Journal* in which I pointed out that the magnetic field of the Galaxy, lying along the galactic arm, is necessarily confined by the weight of the interstellar gas so that there is a limit to the strength of field that can be confined by any given gas density and scale height. I also pointed out that the confinement is dynamically unstable, all based on rather elementary physical principles and extensively illustrated with formal calculations. My office was about 60 ft from Chandra's office and the office of the secretary for *The Astrophysical Journal* was about half-way between. One afternoon the secretary appeared in my office saying that the referee's report on my paper had just arrived and she thought I would like to see it. I thanked her, and since I was doing nothing of importance at the time I read the report. The report began with "I had always thought that Parker was competent, but ...". I read through the report with Chandra's advice in mind, and it was clear that my introductory remarks had to be sharpened up, with some additional clarification at specific points in the text. I had the revision just about completed a couple of hours later when Chandra appeared in an agitated state. He asked me if the secretary had given me the referee's report and I replied in the affirmative. He then explained that the secretary should have given the report to him first for proper editing. I told him it was nothing to be concerned about and that I had nearly finished with the revisions based on the referee's inability to grasp the rather simple physics of the paper. I reminded Chandra of his own advice on the importance of referee's reports, no matter how badly written. But Chandra remained apologetic for his "blunder" until I finally pointed out to him that the paper was really based on some very simple physical considerations. So simple in fact that I had begun to feel that the point of the paper bordered on the trivial. On the other hand, the complete inability of the "distinguished" referee to grasp the point emphasized the nontrivial character of my paper. So I was sincerely pleased with the episode. Chandra cheered up after that and assured me that he did not approve of offensive remarks in referee reports, and I assured him that the offensive remarks were the best part of the report so far as my own frame of mind was concerned.

Chandra once told me a story of a conversation that he had with his famous and autocratic uncle C.V. Raman. Chandra was thinking about writing his book on radiative transfer, and, when he mentioned the fact to Raman, Raman replied that he had once considered writing a book on spectroscopy. However, on reflecting upon the enormous time and effort the writing would require, Raman said that he decided to invest the time in research instead. The research led to his discovery of the Raman effect, for which he was awarded the Nobel Prize in 1930. Clearly, Raman thought poorly of Chandra's idea of a book on radiative transfer. In jest Chandra replied something to the effect that "Then I must have missed two Nobel Prizes because I have written two books." Raman was furious.

As we all know, Chandra's book on radiative transfer has become a classic, although it earned no Nobel Prize. He was awarded the Nobel Prize for his work on relativistic degenerate gases and the resulting mass limit for a star that ultimately fades out as a cooling white dwarf rather than collapsing to a black hole.

One morning not long after we came to Chicago I met Chandra as I was walking north on Ellis Ave. He showed high spirits and as we met he said, "Well, Parker, I have been immortalized." To my puzzled looks he added, "Dover (Publications, New York) has decided to publish my book on radiative transfer." As we all know, Dover went on to reprint most of Chandra's monographs, so Chandra has been immortalized several times over.

Reminiscing About Chandra's Research

Martin Rees

After his pioneering student insights into how stars end their lives, and his discouragement by Eddington, Chandra shifted to other topics. His achievements in the next three decades were systematised in his series of classic texts on stellar structure, dynamics of stellar systems, fluid mechanics, and other specialised topics. But he returned to the study of black holes in much later life.

The early 1970s were the "heroic age" of black hole research. Theorists discovered that, if Einstein was right, black holes weren't infinitely diverse but were standardised objects, characterised just as surely as any elementary particle by their mass and their spin. And astronomers were starting to suspect that black holes were not just theoretical constructs, but might actually exist in our universe.

This made a deep impression on Chandra, aesthetically as well as scientifically. In a lecture in 1975 he said, "In my entire scientific life ... the most shattering experience has been the realisation that an exact solution of Einstein's equations of general relativity, discovered by the New Zealand mathematician Roy Kerr, provides the absolutely exact representation of untold numbers of massive black holes that populate the Universe. This 'shuddering before the beautiful', this incredible fact that a discovery motivated by a search after the beautiful in mathematics should find its exact replica in Nature, persuades me to say that beauty is that to which the human mind responds at its deepest and most profound."

Chandra was already in his sixties when he embarked on black hole research. He was fond of quoting the great physicist Lord Rayleigh's response to T.H. Huxley's claim that 'scientists over 60 do more harm than good'. Rayleigh (aged 67 at the time) had responded: "That may be, if he undertakes to criticise the work of younger men, but I do not see why it need be so if he sticks to the things he is conversant with." It is a precept that Chandra himself manifestly followed. He never fully absorbed the mathematical techniques introduced by Roger Penrose, which had given the subject such impetus. Instead, he made his distinctive contribution by adapting the more "classical" methods he had used in other contexts.

He analysed how black holes would respond when their equilibrium was perturbed, extending techniques that had traditionally been used to study the vibration modes of a drum, or of earth and oceans. The techniques, in a sense, complement the methods which Penrose pioneered: they cannot handle "generic" collapse, where there is no special degree of symmetry; but they yield a more quantitative picture of what would happen if a black hole were perturbed (by, for instance, a smaller object falling into it or orbiting close to it). These techniques offer a "probe" for black holes, just as seismologists can learn about the Earth's structure from the various modes of oscillation when its crust is set "ringing" after an earthquake.

Chandra was unique among his contemporaries in his intellectual stamina, which, combined with his self-disciplined neatness, allowed him to carry through the most elaborate mathematical manipulations without flagging and (equally remarkably) without mistakes. I recall the first time I heard him lecture, at a Cambridge seminar. He presented his mathematics on slides, which he ran through at bewildering speed because each equation was too long to fit on a single slide, and spilled over onto several. He ended his talk with a typical disclaimer: "You may think I have used a hammer to crack eggs, but I have cracked eggs."

Chandra's mathematical virtuosity is dauntingly manifest in his 650-page treatise on "The Mathematical Theory of Black Holes". In one chapter, 100 pages long, the manipulations are so heavy and the argument so terse that he adds the following footnote: "The reductions that are necessary to go from one step to another [in this chapter] are often very elaborate and, on occasion, may require as many as ten, twenty, or even fifty pages. In the event that some reader may wish to undertake a careful scrutiny of the entire development, the author's derivations (in some 600 legal-sized pages and in six additional notebooks) have been deposited in the Joseph Regenstein Library of the University of Chicago."

Such is the aura surrounding Chandra and his subject that this formidable and recondite text has notched up several thousand paperback sales. Its sales are not, of course, in the class of Hawking's *Brief History of Time*, but it has probably surpassed Hawking's book in the ratio of copies sold to copies actually read. Any reader who perseveres would echo the 19th century scholar William Whewell's reaction to the mathematics of Newton's Principia: "We feel as when we are in an ancient armoury where the weapons are of gigantic size; ... we marvel what manner of men they were who could use as weapons what we can scarcely lift as a burden."

Chandra was already 72 when his book on black holes appeared. Most of us suspected that it would be his final monograph — indeed, it rounded off his career with a fitting symmetry, by codifying our understanding of the objects foreshadowed by what he had done as a student in Cambridge, 50 years before (an initial insight which gained him as much acclaim as the ensuing 50 years of intellectual toil). But he continued his unflagging output of highly technical papers.

And he developed a new enthusiasm. His lifelong fascination with individuals who scale the supreme peaks of creativity, whether in science or the arts, led him to a detailed study of Newton's work, which culminated in a 600-page exposition, *Newton's Principia for the Common Reader*, published in 1995. I fondly recall my last meeting with him in March of that year, when he had come to England to check the final proofs at the Oxford University Press. He enthused about the unique depths of Newton's insight; he also, less characteristically (or so it seemed to me), spoke of his plans for a touring holiday in Wales later that year.

Chandra intended his book on the Principia to be his final one. Indeed he may well have decided, at the age of 84, to bring his extraordinarily sustained scientific efforts to a close. He was always critical of elderly scientists who "lived on their reputation"; a clean break was better than the risk of compromising his standards. As he told his colleagues, "there is a time for all things, and a time to end all things". His last paper in the Proceedings of the Royal Society appeared in August 1995, the month he died. Whether he would ever actually have relaxed his arduous regime, and made a clean break from a long life of thinking, learning and working, we shall never know.

As they grow older, some scientists cease doing research. Others (and Eddington should perhaps be classed among them) retain their urge to understand the world, but no longer derive satisfaction from "routine" problems; such people overreach themselves, often embarrassingly, by tackling, and even claiming to solve, fundamental problems that are unripe for solution beyond their real expertise. Neither of these paths tempted Chandra.

Chandra probably thought longer and deeper about our universe than anyone since Einstein. His non-technical lectures, always carefully crafted, were all too rare, but one of the last of these perhaps conveys the inspiration that drove him on. In it he concluded:

"The pursuit of science has often been compared to the scaling of mountains, high and not so high. But who amongst us can hope, even in imagination, to scale the Everest [and reach its summit] when the sky is blue and the air is still, and in the stillness of the air survey the entire Himalayan range in the dazzling white of the snow stretching to infinity? None of us can hope for a comparable vision of nature and of the universe around us. But there is nothing mean or lowly in standing in the valley below and awaiting the Sun to rise over Kanchanjanga."

Reminiscences About Chandra

Robert G. Sachs

The first time I met Chandra was in the period 1939–41, when I was a postdoc working with Edward Teller at George Washington University. Teller and George Gamow used an office of which I was in effect the sole occupant. Each of them would drop in from time to time before teaching a class or to pick up mail between appointments elsewhere.

It was during the period when Gamow's original push into nuclear astrophysics was flourishing and he was using Chandra as the source of all wisdom in theoretical astrophysics. The two of them must have been passing through the office on one of those mail stops between one important appointment and another when we met in the office, because I recall it as a very brief encounter. I doubt that Chandra would have remembered the encounter at all.

Although I had not, and have not, done any work in astrophysics, I knew that I had just met one of the Great Men in the field. Gamow had often spoken of Chandra in those terms in my presence before that meeting. Chandra did appear to be very young for a Great Man but that was no surprise for those of us who were just beginning careers in theoretical physics; almost everyone who was active in theoretical physics at that time was young.

The Ballistic Research Laboratory

It was during World War II, after I arrived at the Ballistic Research Laboratory, Aberdeen Proving Ground, Maryland, in 1943 that Chandra and I really came to know each other. He was among the group of distinguished scientists who worked there as a scientific resource for the Associate Director of the Laboratory, Robert H. Kent. The group included Oswald Veblen, Joseph E. Mayer, L.H. Thomas, and Ronald Gurney. Kent was an unsung hero of WW II who had served in the U. S. Ordnance Department since WW I and had conceived the scientific and technical capabilities which were realized in the formation of BRL at the beginning of WW II. Although the Director of the Laboratory was necessarily a military officer (Col. Leslie Simon) Kent was the *de facto* technical director.

It was my good fortune that, by chance, Chandra (who arrived at about the same time as I did) and I were assigned desks in the same room when we first arrived at BRL. That was how we came to know each other and become good friends.

I know very little about the specific work he did there because it was not directly related to my work except in that all of us were working on problems of terminal ballistics, which included work on shock waves and bomb blasts. It also included experimental work using flash X-radiography to study small explosions in a test chamber that was in a laboratory situated directly under our office.

Those experiments had a very direct impact on Chandra and me because every test-firing made us jump out of our seats. It was not a situation conducive to the contemplative life, but Chandra took it in stride with good humor.

Although we rarely talked to each other about science we did have relaxed moments of conversation about general subjects, and they were delightful moments for me. I learned then that Chandra was fascinated with the characters and foibles of the outstanding players in the story of modern physics. It was from him that I first heard some of the well-known amusing stories demonstrating their eccentricities, especially those of the great British physicists.

He also spoke fondly of aspects of the British world view that he found to be amusing. He wrapped it all up in one delightful anecdote:

There was a period of unusually dense fog extending over the British Isles and the Channel while he was at Cambridge and it continued for some days, disrupting transportation generally and especially transportation by sea. After the first day or so this situation was announced by a London newspaper with a huge black headline: "CONTINENT ISOLATED!"

We also had conversations about social conditions in the United States. Of course he was particularly sensitive about the ambient prejudice against people of color. In this regard Maryland was not typical of either the Southern or Northern states but, at that time, in the eastern rural areas like that encompassing Aberdeen, strong prejudice was prevalent. Although the Army established strong and, ultimately, effective measures to overcome the mind-set of segregation, the message had not come through to many members of the officer corps at Aberdeen as early as 1943.

An illustration of the social atmosphere was displayed in the washrooms of the BRL building, which was quite new. The mirror over one of the sinks in the men's room had engraved into it the word "colored". For the women there were two separate washrooms, one of which was marked "colored" on the outer door. This was a federal building!

There were a number of young women "mathematicians" in the Laboratory using hand calculators to do the myriads of calculations required in ballistics. They had been recruited in two different groups. One group came from southern colleges, the other from northern colleges (mostly New York, I believe). As far as I can recall, both groups included both black and white members. A noisy controversy over the washroom situation arose and spread through the Laboratory because not only were the black women forbidden to enter the unmarked washroom but the white women were also forbidden to enter the one marked "colored". There were resentments in every possible direction. As I recall, the controversy was settled by a "gentlemen's agreement" (so described!) to the effect that the label would be removed but the black women would voluntarily refrain from using the (originally) unmarked washroom. I don't remember whether the sign over the sink in the men's room was removed or not but I suspect that it was.

Imagine what an impression this whole thing made on Chandra! He had told me in response to my question that he commuted from Chicago because he wouldn't think of bringing Lalitha to Aberdeen with him. He said that even if she always dressed in a sari she would not be protected from insults in that social climate.

Wali tells of some of Chandra's unpleasant experiences along this line when he first arrived at BRL.

The Wisconsin Years

After the War, when Chandra was again spending full time at Yerkes, in Williams Bay, Wisconsin, he made occasional visits to the Astronomy Department at the University of Wisconsin in Madison and that gave us the opportunity to get together again since I was in the Physics Department in Madison. I remember two occasions when he actually came to me for suggestions concerning physics but, I regret to say, I was of no help to him.

On one such occasion he sought information on the Lorentz transformation properties of scattering cross sections. The answer to his question would be viewed as trivial now; every experimentalist in high energy physics knows how to boost a cross section from the laboratory frame to the center-of-momentum frame by a Lorentz transformation, but, at the time he asked the question, the development of the manifestly covariant methods required to deal with relativistic phenomena was just getting under way so that I was unable to give him a quick answer. I do not know for what specific problem he needed this information nor do I know whether he resolved the problem, but suspect that he did because, once he started on a problem, he always pursued it to the end.

On the other occasion he raised questions about the calculations by Hylleraas of the ionization potential of the negative H ion. The ion was known to be stable but with a very small ionization potential. The iteration method used by Hylleraas did not converge; the sign of the energy calculated in each order of approximation reversed from order to order. Chandra wanted to know if I could suggest a better approximation method but I could not. In this case it is well known that he developed one himself.

I think that it was on one of these visits that we had a conversation about the restriction in Nobel's will precluding an award in the field of astronomy. Chandra was quite incensed that Bethe's identification of the nuclear reactions responsible for energy production by the Sun had not (at that time) won the Nobel Prize. In our discussion he made an interesting statement about the awards: "It is not that I believe that any of the recipients of the awards are undeserving, it is that I believe that there are some who are more deserving but are not recognized." I am sure that, at the time, the possibility that he would ever receive a Nobel Prize was the furthermost thing from his mind.

The Chicago Years

Beginning in 1952 Chandra started to shift his center of activity from Yerkes Observatory to the campus of the University of Chicago and he finally moved full time to the campus in 1964. It happens that I also came to Chicago in 1964 so that we were able to see much more of each other after that time. As a result I learned much more about his point of view concerning science and about the enormous contributions he made to the University.

He was a member of the Enrico Fermi Institute (EFI), an interdisciplinary organization all of whose faculty members held joint faculty appointments in at least one conventional teaching department. The departments represented include Astronomy and Astrophysics, Chemistry, Geophysics, Mathematics and Physics. Most members of the EFI are members of the Physics Department and some have joint appointments in Physics and Astronomy. Chandra was included in the latter group.

Chandra's presence in the EFI (which, before Fermi's death, was "The Institute for Nuclear Studies") had a tremendous impact on the direction in which the Institute developed, including the creation of the current large overlap between the Physics and Astronomy Departments. The original interdisciplinary thrust of the Institute was set by the senior faculty including Allison, Fermi, Libby, Joseph Mayer, Maria Goeppert Mayer, Teller, Urey, and Wentzel. The interests of this group were very broad. They included nuclear physics and chemistry, statistical mechanics, cosmic rays, astrophysics, planetary physics and two fields now called "particle physics" and "space physics". All of these fields were ultimately relevant to Chandra's interests so that he was a perfect fit into this setting.

From the beginning the interdisciplinary interests of the Institute were brought to a focus at a very informal seminar called "The Institute Seminar". In the early years it was a research seminar with general discussions of research problems initiated by very brief talks concerning research in progress. This turned out to be very fruitful in bringing to bear the insights from a variety of disciplines on problems raised by the speaker concerning his research. I was not present in many of the seminars during the period when this was the dominant style and therefore do not know specific examples of Chandra's contributions then. However the amount of EFI research activity devoted to astrophysics, magnetohydrodynamics of plasmas in stars and in space, nucleosynthesis, abundance of the elements, and so forth was so great that Chandra must have been an active participant. After all he was the authority on whom everyone leaned to get these fields started.

By the time I was attending regularly, the seminar was serving more as a very high level journal club than as a research seminar. As editor of *The Astrophysical Journal* Chandra was in a particularly good position to report on recent developments in astronomy and astrophysics.

This was in the 60's when the application to astronomy of instrumentation based on WW II developments (such as the application of radar to radioastronomy) was leading to remarkable discoveries. In retrospect it seems that Chandra announced a new discovery or a new interpretation of a recent discovery almost every week. This served to guide us in our recruiting of new faculty because it made clear to all of us that there were new, overlapping opportunities in experimental astrophysics, space physics, theoretical physics and astrophysics, cosmology and general relativity to which our interdisciplinary makeup was very well suited.

Chandra was the only physicist at the University of Chicago working in the field of general relativity at the beginning of this period and he had been pressing for the addition of a relativist to the physics faculty. After considerable effort on his part we successfully acquired a small relativity group. And, because of his presence, as well as the presence of the high energy physics and astrophysics programs, we had little difficulty in attracting the first rate, young experimental and theoretical astrophysicists who have established a very strong astrophysics and cosmology program at the University of Chicago.

Some Impressions

Chandra had a great interest in the scientists who had played an important role in the development of physics and astrophysics. He had an impressive grasp of not only the scientific accomplishments, but also the personal histories of many of these people. This first manifested itself to me when he showed his knowledge of the characters and foibles of the outstanding players in the story of modern physics, as I mentioned earlier (in the BRL days).

But it was not only the participants in modern physics who interested him, as can be seen from his discussion of Kepler's motivations and work in the lecture "The Pursuit of Science: Its Motivations" and from the studies of Newton and the Principia that he took up in recent years.

My impression is that he found the creative leaps of such people as Kepler, Newton and Einstein to be so remarkable that he tended to place them on pedestals well above mere mortals (including himself.) During the period when he was working on problems of general relativity he often gave the impression that, in his opinion, Einstein had the greatest and most original mind among all physicists, living or dead. Later, after he had started studying Newton's Principia, he explicitly replaced Einstein by Newton in that category.

He made a point in his conversations with me and in his lectures about Newton that he was astonished at Newton's ability to realize the laws of mechanics and prove their validity using Euclidean geometry as the only mathematical tool. Before reading Newton's proof of an assertion of a law or a theorem of mechanics Chandra would try to prove it himself. He found it easy to do so by using standard modern mathematical methods, but methods based purely on the axioms and theorems of Euclidean geometry did not come naturally to him. He was tremendously impressed by the geometrical insight that led to Newton's elegant proofs.

Chandra also made comments to me concerning his views on the question: "What defines a correct theory or a good theory in physics?" In his "Einstein years" he told me that a good theory is characterized by the simplicity and the beauty of the mathematics. This has much the same flavor as Einstein's statement in a letter to Lanczos describing his (Einstein's) conversion from the views of E. Mach to a more ethereal view of physics. Einstein said: "Coming from skeptical empiricism of somewhat the kind of Mach's, I was made, by the problem of gravitation, into a believing rationalist, that is, one who seeks the only trustworthy source of truth in mathematical simplicity."

Chandra surprised me a few years ago by making the remark in passing: "A good theory is one that has an exact solution." I did not have the opportunity to explore the meaning of the statement, in particular, whether he meant it to be consistent with the notion of beautiful mathematics or to be a distinctly different viewpoint.

In this respect I think it is interesting to go back to Chandra's writings concerning his own motivations for his work. In the preface to his collection of lectures entitled "Truth and Beauty" Chandra describes some of his difficulties in pinning down the aesthetic and other motivations of scientists because of his perception that they have changed markedly over the years. He indicates that these changes are illustrated in a comparison of the first two lectures presented in this collection.

In the first (1946) lecture, entitled "The Scientist", he describes his own motivation in the following words: "The method I have adopted in my own work has always been first to learn what is already known about a subject; then to see if it conforms to those standards of rigor, logical ordering, and completion one has a right to ask; and, if it does not, to set about doing it. The motivation has always been systematization based on scholarship."

The accuracy of this early description of the style of his life's work is remarkable. In recent years he described the way in which he set his agenda in very similar terms. He said that he took a subject that had come up in his work but for which he could not find an adequate treatment and wrote a book on it. When he finished that book, he turned to the next such subject on his agenda for his next book. Each such book became a complete and authoritative treatment of its subject and, as a glance at his bibliography shows, it is a remarkable collection of diverse subjects in physics and astrophysics.

The second (1985) lecture, mentioned above as my source for his discussion of Kepler's motivations, includes anecdotes illustrating the concepts of other great physicists about their motivations and inspirations that led to their important discoveries. There is not a word about Chandra's own work. He seems to have viewed the accomplishments of others with great awe, well beyond his own capabilities. That interpretation is consistent with my earlier observation that he placed certain scientists on a pedestal.

Whether he realized it or not there are many of us who place him on a pedestal for his remarkable contributions, both direct and indirect, to our understanding of astrophysics and cosmology. And his position there is permanent.

Notes

1. Wali, K. C., 1991. *Chandra* (Chicago: University of Chicago Press), pp. 194–195.
2. S. Chandrasekhar, 1987. *Truth and Beauty* (Chicago: University of Chicago Press), pp. 15–28.
3. Chandrasekhar, S., 1995. *Newton's Principia for the Common Reader* (Oxford: Clarendon Press).
4. I have lifted this quotation from Gerald Holton, 1993, *Science and Anti-Science* (Cambridge, MA: Harvard University Press), pp. 65–66. Note 2, pp. vii–x. Note 2, pp. 1–14.

Chandra and Isaac Newton

Stephen M. Stigler

Isaac Newton's *Philosophiae Naturalis Principia Mathematica* celebrated its 300th birthday in 1987. In the fall of 1986, I met with fellow members of the Fishbein Center, an interdisciplinary group of University of Chicago scholars interested in the history of science and medicine, to plan the next year's colloquia. Of course we could not let the anniversary of the greatest single work in the history of science pass by without an appropriate salute. A single day — or even a three-day meeting — would not do. We decided to hold a quarter-long series of talks. There would be no shortage of possible speakers: a host of distinguished historians of science and the editors of Newton's papers on mathematics and on optics came quickly to mind. But we needed more, we needed an initial speaker of sufficient scientific breadth and stature to set the proper tone, so the quarter would truly be a celebration of Newton's signal achievement: a celebration of Newton's science and not merely a celebration of Newtonian studies.

I suggested we invite our own Professor Chandrasekhar to inaugurate the series; all members of the Center thought this was an attractive possibility. Chandra was the University's pre-eminent physical scientist, but the question arose: while his own work certainly followed Newtonian paths, we were not aware of any expression of interest on his part in the history of science, much less in Newton in particular. Still, could someone who was as fascinated with artistic creativity, with Shakespeare and Beethoven as Chandra was, possibly be blind to the charms of Newton and the Principia?

I was deputized to approach Chandra with our proposition. In my initial phone call, I described our series, explaining that we hoped he could speak on the importance of Newton and the Principia to modern work, and to his own work in particular, but that the way he would approach the subject would of course be up to him. His reply was polite, but noncommittal. He would think about it. Two weeks later I reached him for a follow-up conversation. His response was exhilarating to me: not only would he accept the invitation, but in the intervening time he had actually been reading Newton, and his excitement about what he had found was palpable. Of course he had known about the Principia for many, many years — both he and Newton were sons of Cambridge — but he had never before paused to read seriously in Newton's masterpiece. And once he did, there was no restraining him.

Chandra's lecture, inaugurating the celebration of the Principia in its 300th year, was set for April 15, 1987, with a reception following. My memory of that lecture remains vivid; I subsequently heard Chandra speak on Newton on at least three other occasions, but that first had a magic quality. Scheduled to talk at 3:30 PM for one hour, he spoke for two, holding the audience in such rapt attention that we were as oblivious to the passage of time as he was.

Chandra explained that his approach to Newton was to locate several of Newton's major propositions, such as Newton's proof in Proposition LXXVI of the inverse square law of attraction and, as I recall, four others. In each case he would take the proposition as a starting point, but without looking at Newton's proof of the proposition he would carefully construct his own proof, using all the arsenal of mathematical techniques available to him — the fruits of three centuries of development of mathematical physics and celestial mechanics. Only then, with his own proof in hand, would Chandra turn back to the Principia to a careful study of Newton's own demonstration.

Chandra's eyes were glowing as he led us through example after example: his own proof, to show us how really difficult the matter was, even after 300 years, and then Newton's. And in each case, as he emphatically told us, Newton's was better! Through his eyes we could see Newton in a new light. The simple geometric demonstrations of the Principia became magic; he showed us just how delicate they really were, how the elegant simplicity would not come naturally to any but a Newton, how the mathematical constructions that appeared in the proof depended on such subtle mathematical balances that it was truly astonishing that one individual could have derived them. The slightest change in a diagram, or in a construction, and the proof would not work at all; but with Newton's choices the proofs performed with a sublime beauty that Shakespeare or Beethoven would have envied.

Chandra gave other versions of that talk later in the year — one in Washington, one at the Royal Society of London, one at Madras. And he gave others at the University, including one to our Library Society and several to the Mathematics and Astronomy Departments. In every one of these I heard, the same glow was in his eyes as he took the audience gently through an intricate Newtonian construction. His admiration for Newton seemed boundless, and the audience was carried along with him. I believe that it was in newspaper coverage of his 1987 talk in Washington that a reporter asked him how Newton compared with Einstein. Chandra put Newton at the pinnacle of scientific attainment. Einstein was an excellent scientist, although perhaps a few others could have done the same work. But what Newton had done, Chandra said, was on another plane; Chandra could not conceive of himself having done the same. At Madras in December he spoke to a meeting commemorating the centenary of the birth of Ramanujan. To the surprise of some he spoke not about Ramanujan, but about Newton. The mathematician Richard Askey tells me Chandra explained to the audience that while it was indeed true Ramanujan was a great mathematician, we need to keep in mind that there were even greater ones. He chose to honor Ramanujan by linking him with Newton, albeit in a secondary role, in the pantheon of human possibility.

Over the next few years I had a few fleeting hallway conversations with Chandra. I learned that his interest in Newton was not abating, that he had gone back to the Principia and begun a systematic study with an eye towards writing a book on it, the book that would become his *Newton's Principia for the Common Reader* (Oxford, 1995). His talks about Newton continued: he gave a series of lectures at Penn State that was videotaped, and a lecture to a winter meeting of the American Mathematical Society in Cincinnati. One day in mid-October 1991 I ran into him in Eckhart Hall, and he stopped briefly to chat. I asked him how his work on Newton was proceeding, and I added: "You know, the marketplace is putting an increasingly high value on Newton, too. I have just received a catalog from a rare book dealer offering a first edition of the Principia for L50,000!" He nodded, told me of his progress, of his still increasing admiration for Newton, and of his contempt for much of modern scholarship on Newton, by people who did not read the Principia and whose emphasis of social factors in the history of science frequently masked their lack of understanding of the science. We parted. An hour later I was sitting at my desk, when the phone rang. It was Chandra. "Did you say," he asked, "that you knew of a first edition of the Principia for sale for $5,000?" I had to disappoint him about the price, but I offered to follow up, since I know many dealers in antiquarian books, and see if perhaps another copy was available for a lower price. After all, I told him, I recalled a copy available in 1987 for $42,000, and one only a bit earlier for $25,000.

On October 23 I faxed a question to Andrew Hunter in London, the head of the science department at Bernard Quaritch Limited, one of the world's oldest and largest antiquarian book dealers, and the most likely source for a Principia. "Do you presently have a copy of the first edition of *Newton's Principia*? Do you know of a copy?" Hunter's response was not encouraging. The copy that had been offered for L50,000 by another dealer had in fact been jointly owned by Quaritch; furthermore it had sold! The sharp rise in price, he wrote, reflected an increase in demand that was far from satisfied. He told me of other recent sales at auction for $77,500, $55,000, and $105,000, all exclusive of the 10% auction house premium, and of a copy that had sold immediately after being listed in a catalog for $120,000 by a New York dealer, Jonathan Hill. The only encouraging note was his parting "We do have a nice copy of the second edition..."

I replied: "Thank you for the Newton information. At those prices, I expect the supply will soon rival the print run!" I also asked for their listing on the second edition, that of 1713 and known to scholars as the Cotes Edition, after its editor, Roger Cotes. That listing arrived October 28; it described a handsome volume which included, bound in, a bonus: the 1732 Leiden edition of Newton's *Arithmetica Universalis.* And best of all, the price was less than a tenth of the price of a first edition.

I took the bad news about the first edition and the listing for the second to Chandra. After brief consideration, he decided on November 3 that he would purchase the second edition, and I placed the order to Quaritch. The book arrived on November 6. I recall that day vividly. It was unseasonably cold; there was a dusting of snow and a strong, cold wind from the west. I bundled up Newton and set out to take it to Chandra, who lived around the corner from me. He received me gracefully and with some excitement, and for an hour we went over the book together. Again his eyes glowed as he showed me one after another of Newton's magnificent proofs. And he showed me where a reversal of placement of two diagrams on one page had led subsequent editions to make the grievous error of "correcting" the text to make it consistent with the diagrams, at the expense of making nonsense of the demonstration. Chandra did not place much value on material possessions, but he once described this book as the nicest thing he owned.

Chandra's work on the Newton volume continued steadily, as I learned from our occasional meetings. Once he stated his worry that he might not live to complete it, but I suspect that if anything, Newton energized him and prolonged his life. One evening — I believe it was in 1993 — he called me, improbably, to ask for help. In the last proposition of Book III (Proposition XLII, Problem XXII), Newton had written of how to determine a comet's orbit from observations. Chandra asked if I would take a look at the third operation and see if there was a statistical aspect to the derivation that would help clarify what Newton had done. I was more than a little doubtful that I could offer anything, but I took a careful look at Newton over several hours. I was right — the passage in question was not at all statistical; it calculated an orbit from three observations using a first order local approximation after a transformation. I went to Chandra the next day to report this. Of course, by then he had discovered the same thing, and much more, but he was quite polite in explaining this to me.

In February 1994, as his book was nearing completion, Chandra had one final request for me. For many years his desk had been graced by a copy of the bust of Ramanujan that Richard Askey had had made, a bust based upon Chandra's copy of Ramanujan's passport picture. Chandra planned to give this bust to the Royal Society of London in May of that year, and he worried about the blank space it would leave on his desk. Perhaps, he asked, I could locate a bust of Newton to take its place? Again I put the question to Andrew Hunter of Quaritch. He replied that, while they did not have a bust of Newton, he would put out feelers in the Trade. He did allow that he knew of an eighteenth century bust "which is hugely expensive." To my question of how much "hugely expensive" might be (I was under no illusions — I knew I was dealing with a firm that routinely handled $100,000 books), he replied that it was L500,000! Newton in marble was more than 10 times even a first edition of Newton in paper. I did not think that would appeal to Chandra. I replied to Hunter that "for that price I would expect to get a personal visit from Newton himself!" We kept searching, but could turn up nothing better than a 1731 pewter portrait medallion at L650 and a two-and-a-half-foot-high nineteenth century spelter ("the poor man's bronze") statuette at L5,500, neither of which would do.

I suspect this was all for the best. Chandra did not think much of most portraits of Newton, and I doubt that we could have found an affordable bust that would have passed his scrutiny. He told me that the only portrait of Newton he liked was one from the National Portrait Gallery in London showing Newton with a copy of the third edition of the Principia open to pages 104–105; he planned to use that in the book. That portrait, executed in 1726 when Newton was 83 years old, does appear in Chandra's book as Plate 2 on page 300. It is in fact a rather dull and lifeless portrait, and according to Derek Gjertsen (*The Newton Handbook*, Routledge, 1986, p. 441) it is now believed to be a copy. I suspect that it is the rendition of the Principia in the portrait that caught Chandra's eye, rather than that of Newton! The only other portraits of Newton in Chandra's book, on the cover and on the frontispiece, are of a bust and a statue, both from Trinity College, Cambridge. For Chandra the real portrait of Newton was in the words and mathematics of the Principia. The marble was attractive, but lifeless; in the book Newton not only lived but spoke to Chandra in a common language.

Chandrasekhar's Research on Newton's *Principia*

N.M. Swerdlow

Chandra was scheduled to give a talk on Newton's Demonstration of Universal Gravitation at the Symposium on Isaac Newton's Natural Philosophy, held on November 9–11, 1995, at the Dibner Institute for the History of Science and Technology at the Massachusetts Institute of Technology. Unfortunately, he suffered a heart attack and died suddenly on August 21, 1995. There was no way of replacing the lecture he would have given; no one knew the subject as he did. Instead Jed Buchwald asked me to say a few words about Chandra and his work on Newton, knowing that Chandra had discussed his study of the *Principia* with me during the years it was in progress. I also had the good fortune of knowing him over the past two decades. I accepted the task with much humility and still with more regret at what was lost with the passing of one of the great scientists of our age. The ensuing article is based on that talk.

In 1969 Chandra published a book, *Ellipsoidal Figures of Equilibrium*, on the conditions for stability and instability of rotating homogeneous fluid masses. The subject is pertinent to stars, systems of stars, gaseous clouds, and more, but here given an austerely mathematical treatment with hardly a word on its application. Chandra could well have quoted Newton's remark in the introduction to Section 11 of Book I of the *Principia*: "But now we are concerned with mathematics, and therefore, setting aside physical interpretations, we use the customary language by which we may be more easily understood by mathematical readers." As was Chandra's practice, the book was preceded by about forty articles published over a period of eight years, in this case a number of them with Norman Lebovitz of the The University of Chicago's Mathematics Department, and by four Silliman Memorial Lectures given at Yale in 1963. When the series of articles and the book were close to completion Chandra gave a colloquium at Yerkes; it was in fact one of the Yerkes centennial colloquia, with a number ending in two zeros, which Chandra regularly gave, in this case number 700 or 800. One might ask, he began, why one should work on so old a subject. His answer had nothing to do with present-day applications. In a historical introduction he reviewed the earlier researches beginning with Newton on the figure of the earth and going on to Maclaurin, Jacobi, Dirichlet, Dedekind, Poincaré, Roche, G.H. Darwin, and Cartan, remarking that "it is such a pleasure to work on a subject that has not been cluttered by mediocrity." The only one in the audience who laughed was Chandra's wife, Lalitha, and afterwards Chandra told Peter Vandervoort, to whom I owe this story, that "my colleagues have no sense of humor." Chandra had a notable sense of humor, and it was most often in the form of understatement. The title *Newton's Principia for the Common Reader*, a phrase borrowed from Samuel Johnson, is an example. Chandra had long had an interest in the history of physics in general and Newton in particular. In 1975 he gave the second Ryerson Lecture at The University of Chicago on "Shakespeare, Newton, and Beethoven, or Patterns of Creativity," although in truth Newton is there passed over nearly as briefly as Beethoven and, perhaps in keeping with a talk for a diverse audience, more attention is given to Shakespeare. In 1979 he lectured on E.A. Milne at Oxford and in 1982 gave two Eddington Centenery Lectures at Trinity College, Cambridge; both played important, and in part difficult, roles in the early part of Chandra's career. It was not until he was asked to give some lectures for the three-hundredth anniversary of the publication of the *Principia* in 1987 that he began to read the book in a systematic way. His method was indirect, but highly educational. He would first read the enunciation of a proposition, grasp its point, and then close the book and prove the proposition in his own way, meaning according to the methods of classical mechanics as treated at the present day. Next he opened the book and studied Newton's own demonstration. What he found again and again was that Newton's proof was superior and showed what he called "deeper understanding" in the sense of getting at the underlying physics in the most direct way. The exposition in the book itself is based upon this same procedure in order that, as he put it, "the physical insight and mathematical craftsmanship that invariably illuminate Newton's proofs come sharply into focus." An example occurs in the very first proposition of Book I, the law of areas. In a preliminary consideration of the motion of particles under centripetal forces, Chandra proves the law of areas, as it commonly is in modern physics texts, as a consequence of the conservation of angular momentum, a perfectly good if indirect proof from a principle of the variation of velocity with distance from the center of force. In the next chapter, he gives Newton's proof in which the equable description of areas is an immediate deduction from nothing more than the first two laws of motion, the parallelogram of forces, and some very elementary geometry of the equality of triangles. Likewise, for Proposition 71, on the attraction of a spherical shell on an external particle, he first gives a proof that J.E. Littlewood conjectured Newton must originally have used, based upon taking circular sections of the shell normal to the direction from the particle to the center, and then shows Newton's far more surprising proof taking the particle at different distances from equal small arcs of a great circle on the shell, which, by rotation about the diameter of the sphere, become precisely the circular sections used by Littlewood. A more extended presentation of this method is the treatment of the three-body problem, specifically of Proposition 66 and its 22 corollaries, which Chandra says would better be described as a "monograph." He begins by setting out a concise exposition of the variation of elements and the derivation of the disturbing function, with their implicit application to lunar theory to be taken up in the corollaries. He then goes through the proposition and the corollaries in detail, sometimes anticipating the lunar theory in Book III, showing how Newton's own demonstrations, or in quite a few cases outlines of demonstrations in what he calls Newton's "secretive style," agree or differ from the variational equations. Whether all of the surmises about the analysis that underlies the cryptic corollaries are correct I do not know, but surely no one has more deeply penetrated Newton's "secretive style" than Chandra has here. At the end of the 17th corollary, in a "personal reflection," he compares Newton's intense concentration in laying the foundations of lunar theory to the story in the Mahabharata of Arjuna's intense concentration on but one thing in the entire world, the head of a bird in a distant tree, before shooting it off with an arrow from his bow, "stretched until it stood in a circle." The reflection is indeed "personal," and there is no better description of the concentration with which Chandra himself worked. The application of the corollaries of Proposition 66 is shown in yet greater detail in the exposition of the lunar theory in Book III, aided again by frequent comparisons of Newton's demonstrations with the variational equations. A similar treatment is given to the tides, also based upon corollaries of Proposition 66, and the precession, both of which understandably pose problems in their application. Chandra's study of Newton is not really historical in the usual sense. The brief historical introduction in the first chapter goes little beyond Newton's own remarks on the origin and motivation of his work. Predecessors and contemporaries play almost no role, which is of course very nearly true. The name of Descartes is hardly mentioned, and no remark is made upon the numerous passages of the *Principia* that are present only to distinguish the correct way of philosophizing about nature from the then fashionable, but in Newton's eyes contemptible, Cartesian "fictions." He relied upon the Mott–Cajori translation with consultation of the Latin texts principally to distinguish alterations in the three editions, and trusted his own understanding to correct errors in the translation, that is, what one might call English rather than Germanic philology. Chandra seemed to regard Newton not so much as an object of historical study as his senior colleague, in a way his only senior colleague, from whom he and every serious physicist had much to learn. And perhaps for this reason, the book contains a great deal of direct quotation from the *Principia*, often without any comment, evidently because there was nothing to add to Newton's own words. Chandra did something like this in his own work. "Very often in my books," he said, "when I have a key idea, and have to restate it at a later stage, I don't leave it to chance. I go back and copy exactly what I had written before." Why then leave Newton to chance? Who could express Newton's meaning better than Newton? And for this there was one reason above all others, the originality of Newton himself. Chandra once remarked that when he read the work of older scientists he would ask himself the question, "Could I have done that?" meaning, I presume, not "Could I have asked that question in the first place?" but "Given that particular problem, could I have arrived at the same solution?" His answer, he said, was always yes and that, by the way, included general relativity until he read Newton. Here, he found, were ways of thinking he could not anticipate, solutions he would not have thought of. And nothing could so capture his interest as someone from whom he could learn.

He worked steadily at the *Principia* until late 1994, sending chapters to Oxford as they were completed to his satisfaction so there would be no delay in bringing the book out as soon as it was finished. During these years, he thought and talked of Newton constantly, and gave any number of lectures, including a series of ten in Chicago and Oxford. Although not exhaustive, as he was the first to admit most of Book II is not considered and with whatever idiosyncrasies and imperfections it may have, what he wrote in these few years is the first detailed, technically proficient exposition of the *Principia* in its three-hundred-year history. And it will probably be some years before the full extent and depth of Chandra's exposition is taken into the history of physics. After giving his lectures in Oxford, one auditor told him that he thought it would take him ten years to read what Chandra was writing. (One thinks of Napoleon's remark to Laplace on receiving the first two volumes of the *Mécanique Céleste.*) Shortly after the book's publication last May, two laudatory and intelligent reviews by highly competent physicists appeared in *Nature* and the *Times Supplement of Higher Education.* Chandra was not pleased with either of them. Why? The authors, he said, had obviously not read the entire book, the minimal requirement, one would think, for writing a review. Chandra saw the history of science as a part of living science in which we learn from our predecessors just as we do from our contemporaries. He referred to historical studies as "scientific work," just as he did to contemporary research. Even if we do not use what we learn because certain subjects have been exhausted or entirely changed, a thorough knowledge of a science requires a knowledge of its history; the ideal scientist must also be a historian and the ideal historian a scientist. Necessarily, therefore, the technical and scientific level of historical research has the same standards and requires the same competence as living science. When the history of science is understood in this way, *Newton's Principia for the Common Reader* is more than a landmark in studies of Newton; it is a scientific treatise that marks a division of before and after in our knowledge of the history of physics and celestial mechanics. In this way it does not differ from Chandra's classic treatises *Introduction to the Study of Stellar Structure*, *Principles of Stellar Dynamics*, and *Radiative Transfer*, not to mention *Ellipsoidal Figures of Equilibrium and the Mathematical Theory of Black Holes*, each of which also marked a new foundation upon which research continues to be built to this day. So with the publication of Chandra's work a new foundation has been laid and a new standard set in the study of the *Principia*, a foundation upon which every student of Newton must now build and a standard he must henceforth aspire to meet or exceed.

Recollections About Chandra

V.L. Telegdi

It is both an honor and an embarrassment for me to write about Chandra. It is an honor because although I knew that great scientist closely over a period of some forty years I, an experimental particle physicist, am hardly well qualified to reminisce about him. It is an embarrassment because I lack the literary skills for conveying the lofty feelings that personal contact with him evoked in me (and no doubt in others). Let me surrender to vanity and try nevertheless.

Let me start by comparing Chandra to other great theoretical physicists whom I had the good fortune to meet and know well, namely Pauli, Fermi and Stueckelberg. Pauli, notwithstanding his immense mathematical talents, used mathematics in his lectures sparingly and put the main emphasis on logical consistency in his classes. The physical assumptions that went into the first equation were the essence, and the steps between that equation and the final result were just routine. He had broad interests outside physics, particularly in psychology and epistemology. His legendary acerbic wit could easily be misleading; the veil of sarcasm covered a warm human personality. Fermi's lectures were models of pedagogical skill and used the barest minimum of mathematics — simplicity, not logical rigor, was Fermi's aim in classes. As a person, Fermi was extremely reserved. As opposed to Pauli, he rarely expressed judgements, good or bad, about other physicists or their work. He was deeply interested in all of physics, but displayed little interest in outside matters or, say, in the epistemological consequences of quantum mechanics. He did everything to be perceived as an ordinary person, the proverbial "man in the street". Stueckelberg, or rather E.C.G. Stueckelberg de Breidenbach as he liked to style himself, was the "mad genius", albeit not quite in the sense that lay persons tend to attach to this expression. Like Pauli, he was primarily interested in general problems rather than in applications and downplayed mathematics in his classes. His extreme originality of mind unfortunately extended also to such routine matters as notations — making both his lectures and his papers hard to digest. He was modest and aristocratic at the same time; he showed little pride in his great (and largely overlooked) contributions to physics, but was conscious of his noble origins and his military rank. In his later years, he developed a certain interest in religious mysticism.

Chandra was radically different from all these great men, and I doubt that this difference had anything to do with his non-European origin. His lectures — like his papers — were polished to the extreme, both in respect to language and to mathematical elegance. I once remarked that whatever the official title of any one of his courses was, it should have been "Mathematical Methods of ...". Rather than being interested in new laws of Nature, Chandra strove to produce exact (and in general analytical) solutions to specific problems. In this sense, one might consider him a mathematical (rather than theoretical) physicist, in the great British tradition of Stokes, Rayleigh and Larmor; I am sure that Chandra would not object to this classification. In his contacts with others, Chandra was always reserved, slightly formal and somewhat aloof. I do not think that this was, like in Fermi's case, a strictly personal feature, but a remnant of his Cambridge years. He would rarely volunteer judgments of other scientists or their work, but could occasionally provide, in private and if prodded, quite scathing comments. His modesty was, I believe, also more a matter of style than of character and it never assumed the caricatural proportions displayed by, among others, Eugene Wigner. He was aware of his exceptional gifts and would have placed himself exactly at his objectively deserved rank. He could be fair and generous to a fault. Notwithstanding the fact that Eddington had almost prevented the proper appreciation of Chandra's first great paper, Chandra would gracefully refer to him as to "the greatest astrophysicist of our time". Chandra's strong interests outside his field, i.e. in literature and music, are well documented by his speeches and writings (e.g. his book *Truth and Beauty*). What physicist, other than Chandra, would draw parallels between certain aspects of general relativity and Monet's series paintings?

The reader who has borne with me so far will complain about the absence of specific personal recollections. Let me try to satisfy him with a few true anecdotes. The first of these concerns Chandra's relation to postwar India. When India in 1961 invaded and annexed Goa, I was rather irritated that such an action, which anyway would not solve any of India's rampant problems, would be taken while Pandit Nehru, a well-known, self-styled apostle of peace, was heading their government. I wanted to discuss this matter with Chandra (in particular because he knew Nehru personally!) but was hesitant to do so for fear of alienating him. Somehow I found the courage and the right moment to discuss this delicate matter. To my great surprise, Chandra largely shared my views and suggested that perhaps the action had been taken to deflect the electorate's attention from domestic problems.

The second incident concerns a little improvised talk that Chandra gave in one of the Fermi Institute's weekly seminars. It concerned a certain paradox that Gödel had discovered in general relativity. Something physically unreasonable was supposed to happen when an observer traveled along some funny closed world line. Before Chandra had had a chance to develop his formal argument, I raised my hand and asked: "Perhaps that observer is doing what is forbidden in relativity, namely moving at the velocity of light?" Chandra admitted instantly that this was essentially the solution and held no grudge for my having, without malice, stolen his thunder.

Last but not least, there is the matter of my collaboration with Chandra (science "creates strange bedfellows", you might say). Since my student days in Zurich I had been puzzled by the so-called Sagnac effect about which I had read in Joos's one-volume *Theoretical Physics*. Here an optical interference apparatus is mounted on a turntable and the rotation of that table can be detected by a shift of the interference fringes. Joos made the curious assertion that this effect is "the optical analog of Foucault's pendulum". "What would Mach have had to say about this effect?" I asked myself, since Mach had wanted to attribute the inertia of local matter to the effect of distant stars — but had no interaction between gravitation and light. In general relativity there is, however, such an interaction. I turned with these questions to Chandra, who right away started to calculate, in GR, the effect of a massive shell rotating about Sagnac's turntable, now locally at rest. We had many discussions: while I was stunned by Chandra's flawless mathematics (his scratch sheets were more perfect than the final notes prepared by other mortals), he seemed to be impressed by my ability to guess certain of his answers by physical arguments (generally borrowed from analogies with classical electrodynamics). In the course of this activity, we studied one of Fermi's earliest papers, one in which the plane of polarization of light is dragged along by the rotation of a dielectric cylinder. Chandra quickly rederived Fermi's result, finding even a minor error in it. For reasons which I do not recall precisely, we never published our results. Perhaps we argued like my Swiss friend Markus Fierz, who under similar circumstances once had said: "Why bother to write this up?

We now fully understand the problem and we both have tenured positions." Anyway, this close interaction with Chandra constitutes one of my fondest memories.

It is now generally known that when Chandra was first appointed to Yerkes Observatory his dark color was objectionable to some of the faculty. I did not know this. At one time Chandra and I were discussing some ethnic massacre (in Europe) in the news, and Chandra made the comment "When many thousands of people were killed in Indonesia, nobody seemed to make much fuss about it." I replied: "That is probably because those victims were not white." Chandra was visibly touched by this remark, and said: "Val, you have hit on the right point." Only years later could I understand that he himself had been discriminated against.

In closing this eulogy, I would like to add a minor negative note. Chandra was admired and respected by every member of the Chicago Physics Department, and therefore he could have exerted an immense influence on its decisions. He generally chose not to use his authority to vigorously oppose actions which were not to his liking. I do not know whether this was due to his great reserve in all situations, or to his inclination to take sides only on those issues in which he was an obvious expert. In general, he agreed with my views in appointment matters in private but would not engage in a battle on the floor of the meetings. When I once asked him why my own interventions were rather ineffectual, he simply replied: "Val, people do not like to hear the truth." A sobering, perhaps even cynical, comment from a man who knew both Truth and Beauty.

Chandra's Visits to Poland

Róża Michalska-Trautman and Andrzej Trautman

We met Chandra for the first time in the summer of 1962, during the International Conference on General Relativity and Gravitation, held in Jablonna, near Warsaw. Our closer acquaintance with the Chandrasekhars, which later developed into friendship, started in the early seventies, after A. had spent four months at the University of Chicago, at Chandra's invitation. Since that time, both Chandra and Lalitha visited Poland several times. We remember most vividly their three short stays in our summer cottage in Zielonowo, a small village in Warmia, a region of lakes and forests, about 120 miles north of Warsaw. In Zielonowo, we spent time walking in the woods, bathing in the lake, and on quiet conversations in front of a fireplace. During the first visit there, Chandra enjoyed riding on a bicycle in the company of our sons, Pawel and Krzysztof, then aged 12 and 10, a thing that he had not done for years. Dressed formally in an immaculate dark suit, he liked the simple life and food of the countryside. The Chandrasekhars were easy to entertain. Lalitha often helped in the kitchen and we found it very hard to convince Chandra that it was not his duty to wash the dishes. On the other hand, one had to be rather careful with arranging social gatherings; once or twice we hurt Chandra by imposing on him the company of people that he did not like. Chandra strongly disliked the presence of people who smoked or talked too much. We have the impression that, most of the time, we understood his moods, but sometimes he was angry for a reason that was not quite clear to us.

A few times, we made excursions in the vicinity of Zielonowo. Chandra and Lalitha were very interested in the old wooden buildings gathered from the region in a *skansen*. Chandra was also moved by the sight of an old, steam-engine train. He enjoyed looking at children from a nearby country school and conversed in German with an elderly lady, a native of Zielonowo.

One of our friends described his first impression of Chandra as that of meeting a "superior being". There was something in this, in the sense that people accepted his superiority without feeling humiliated. One day, in Zielonowo, Lalitha said: "Chandra wants to speak with you, Róża." Róża felt like going for a very important audience. (A. was on more familiar terms with Chandra; they spent much time together.) We discussed, besides physics, many problems. Chandra had a good understanding of the difficult situation of Poland in the '70s and '80s and of our aspirations for sovereignty and democracy. Even though he himself favored the Democratic Party over the Republicans, he understood that we preferred Ronald Reagan over the democratic candidates to the American presidency.

Chandra was also interested in Polish physics and astronomy. He knew very well the work of Marian Smoluchowski on fluctuations. He wrote a major paper on applications of Smoluchowski's theory to astrophysics ["Stochastic problems in physics and astronomy", *Rev. Mod. Phys.* **15**, 1–89 (1943)]. His words about Smoluchowski[1] had a special significance, for Polish science, during the dark years of World War II. He also knew and highly valued the work of Czeslaw Bialobrzeski (who was the first to consider the role of radiation pressure in stars) and Wojciech Rubinowicz (who found the selection rules for quadrupole radiation). Leopold Infeld used to say that Chandra was probably the only person who carefully read his book *Motion and Relativity*, written with Jerzy Plebański. In 1973, Chandra gave an invited paper at a symposium devoted to gravitational radiation and gravitational collapse held in Warsaw in connection with the Copernican Meeting of the International Astronomical Union. During the symposium, he was awarded the Marian Smoluchowski Medal of the Polish Physical Society.

Two conversations are memorable. One evening in the spring of 1971, in the apartment at Dorchester Avenue, Chandra said: "Andrzej, you are almost 40 now. This is a critical age for a scientist; if he is moderately successful, then he is likely to be under pressure to assume administrative responsibilities in the academic world, to become chairman, director or head of something

[1]Chandra wrote: "The theory of density fluctuations, as developed by Smoluchowski, represents one of the most outstanding achievements in molecular physics ... and it is somewhat disappointing that the more recent discussions of the laws of thermodynamics contain no relevant references to the investigations of Boltzmann and Smoluchowski. The absence of references, particularly to Smoluchowski, is to be deplored since no one has contributed so much as Smoluchowski to a real clarification of the fundamental issues involved."

or other. These duties are important for the normal functioning of academic institutions, but they also severely restrict the research activity of the persons who undertake them." A. was later very sorry for not having followed the wisdom of this advice. In Zielonowo, Chandra asked: "Andrzej, why do you do research?" The prompt and sincere answer was: "Because this is what I enjoy; doing science, research and teaching gives me pleasure." Chandra did not like this at all. He said: "For me doing research is an obligation; I do it out of a sense of duty and responsibility towards science and society." But Chandra also enjoyed his research and was very sensitive to the beauty of mathematics and physics; for example, we recall that during his work on black holes, he often expressed amazement and delight at the "miraculous" separation of variables of the equations of mathematical physics when solved on the Kerr background.

Writing about Chandra's visits it is not possible not to mention Lalitha. Her radiant presence, thoughtfulness, vivid interest and participation in conversations contributed, in a significant way, to the atmosphere.

The Chandrasekhars often brought or sent us beautiful gifts. We especially remember the parcel containing a box of English fruitcake that they sent us for a sad Christmas, at the beginning of martial law in Poland, when the situation was particularly grim. In the spring of 1995, R. received from Chandra a newly published biography of Marie Curie by S. Quinn. We planned a reunion in Zielonowo for August of that year. The dates were already set when Chandra telephoned to tell us about the car accident. He said he had not been hurt, but the visit to Poland had to be postponed. His last letter to us was dated July 16, 1995. We were in Zielonowo, preparing the cottage for the visit, when the fatal news was announced on the radio.

When we were invited to write our recollections of contacts with Chandra, we were shy and confused at first. How to write properly about such a great man, a master of the English language who always found an adequate and clear expression for his thoughts? We looked for help from V.S. Naipaul, a writer of Indian origin, who, in *An Area of Darkness*, describes the phenomenon of people like Chandra in the following words:

> ... India produced so many people of grace and beauty, ruled by elaborate courtesy. Producing too much life, it denied the value of life; yet it permitted a unique human development to so many. Nowhere were people so heightened, rounded and individualistic; nowhere did they offer themselves so fully and with such assurance. To know Indians was to take a delight in people as people; every encounter was an adventure.

Meeting Chandra

H.C. van de Hulst

"Giraffe among the poets." This was the opening line of a poem in honor of the Dutch poet Gerrit Achterberg. I like to borrow this metaphor: "Giraffe among the astrophysicists" would be a most fitting name to honor Chandrasekhar. For, with gigantic strides he could rapidly enter a new field and with seeming ease digest the food from branches utterly out of reach to most of us.

Somehow, a line from Keats also keeps popping up in my mind: "The inhuman dearth of noble natures." Perhaps I should avoid citing poetry that I do not fully understand. But if it is read to mean simply that truly noble persons are pitifully scarce in number, I may add that Chandra was one of the few.

I had acquired a deep admiration for Chandra's person and work, both from reading his papers and books, and from personal contacts. However, only recently, when I had finished reading both the biography[1] and the set of reprinted special lectures,[2] did I realize more fully the extent of Chandra's background and erudition. It simply had not occurred to me that he could have been so deeply versed both in fundamental mathematics and in theoretical physics, not even to mention his deep ties with the cultures of two continents.

Chandra was an ardent reader. At one time he told me that during his formative years, he had noticed his tendency to leave a book partially unread, because he had already acquired the next, even more exciting book. At that time — Chandra told me — he made the personal resolve to finish reading at least half the number of pages of any book he bought. He added that he had kept this promise and, as a consequence, had read several books to exactly the middle page. Our last meeting with Chandra was in 1994. He invited my wife and me to join Lalitha and him for dinner at The Hague. He had been invited to be there as one of the Nobel laureates, to act as observer at an international hearing on child labor. Apparently, he approached also this task with his full heart.

In the remaining part of this brief essay, I wish to recount two episodes in which I had a personal taste of Chandra's method of work. They both refer to the area of "Radiative Transfer." To be sure, this area covers only some 15 per cent of Chandra's life-work. But he has also described the years, during which he studied this topic, as the happiest time in his life. Let me explain how I stumbled into this area sideways.

In my wartime study years, I had become interested in radio astronomy, in the interstellar gas, and in the interstellar solid particles (later called grains). With a boldness typical for my age, I was preparing a thesis that was supposed to contain everything known about interstellar matter. One day Oort said to me: "You cite the computations which Greenstein and Schalen have made on the light scattering by the grains on the basis of the Mie theory for light scattering by spherical particles. But would it not be advisable to study this theory yourself?" Acting on this advice, I decided to add a chapter on Mie theory to my thesis. But this expanded into many chapters and finally became my full thesis. I defended it in 1946.[3]

In the meantime, weeks after the cessation of the war in The Netherlands, in May 1945, Gerard Kuiper drove up in a jeep onto the observatory hill in Utrecht to visit Minnaert for an hour. Only much later did we learn that his primary mission had been to be one of a party of physicists trying to find out how far the Germans had come in developing the atom bomb. Minnaert invited a few of his students to join the discussion, and one of the happy side-effects was that I received an invitation to become a postdoc at Yerkes Observatory, which I most happily accepted. A year later, in July 1946, I found myself in a new country, newly married, and with a fresh Ph.D.

The first episode I wish to relate is from those very first days. The Yerkes Observatory is located in a rural setting. In a wide area, adjoining the observatory grounds and the golf course, are the residences of the professors. Chandra and Lalitha lived close to Kuiper's home, where we spent our first month.

I have no distinct memory of my first meeting with Chandra, but quite vividly of an episode of a few days later, when Chandra and Minnaert met. On a fine summer evening, I was sitting in the library reading a book. Minnaert, who had just arrived to deliver a summer course, was scanning the shelves. Then Chandra came in, politely introduced himself and inquired after Minnaert's health and travel. A little while later the conversation had turned to the topic of Minnaert's lectures: spectrophotometry of the planets. Minnaert had shortly before that time invented a method to check the homogeneity of planetary atmospheres by comparing the photometry at two points of the planetary surface, which should be related by the reciprocity principle. Chandra had also met reciprocity in his current work on radiative transfer, so a lively discussion followed. I sat quietly listening, and after a while I had the impression that the two persons, although very polite and eager to understand, did not fully understand each other. Then came the moment I most vividly remember. Minnaert stood with his left hand slanted in the air, like a planetary surface, and one finger of his right hand pointing at it like an incident ray of light. He said: "Let us look at it in the simplest manner." Chandra looked at the finger, but clearly his mind worked differently. For only a minute later I saw him pointing with his finger at the empty table top, as if this was covered with equations, saying: "It is very simple, isn't it? This matrix is symmetric." I was amused and relieved that both experts appealed to simplicity as a quality for better understanding. But they clearly did not have the *same* kind of simplicity in mind. I have often thought back of this conversation and in my own teaching I have made a deliberate effort to show that the same problem can often be approached in different, fully equivalent ways.

The next episode is about a year later. Kuiper, aware of my interest in light scattering, asked me to review for a national symposium on planetary spectroscopy the scattering by planetary atmospheres. While I gladly accepted this assignment as a challenge to enter the field of multiple scattering which was new to me, I was aware of a danger. The fact that this topic was very much the field of Chandra's specialization, combined with the fact that I was not one of Chandra's students, meant that I risked becoming an intruder trespassing on ground where I would not be welcome. I decided to study carefully all twelve papers which Chandra had published so far (the book had not yet appeared) and intended to minimize the risk by highlighting not the theory but the needs and interests of the experimental physicists.

What actually happened did not quite match these intentions. Chandra's papers included elaborate studies of the properties of the X- and Y-functions, initially introduced by Ambartsumian. Chandra treated these functions entirely as mathematical concepts. But the rule that simple mathematics often corresponds to simple physics made me extremely curious to know if these same functions would also have a physical meaning.

This hunch turned out to be correct. In order to explain, I have to become a little more technical. Assume a plane-parallel slab (=cloud layer) consisting of particles scattering light isotropically with an albedo a. Let the optical depth perpendicular to the layer be b. Now place an isotropic light source just above the slab and look at the combined configuration of source + slab from a large distance. Looking down (under some angle arc cos μ with the vertical), we see not only the unimpeded light source above the slab, to which we may assign the intensity 1, but also a blob of multiply scattered light diffusely returning from the atmosphere. The total radiation received in that direction thus is larger than 1 and is called X. This suffices as a "physical definition" of the X-function. This X is a function of a, b, and μ. Similarly, if we look up, making sure that the slab is located between the viewer and the isotropic source, we define the Y-function, again a function of a, b, and μ.

I was excited to find that these simple definitions, when combined with reciprocity, flawlessly led to all the known properties of the X- and Y-functions, including their nonlinear integral equation. Some new expansions also were found. Also, Chandra was pleased. He accepted my intrusion in good grace and my paper went to *The Astrophysical Journal*[4] without problem.

The sequel to this story is sobering and surprising. When many years later the collected works of Ambartsumian had been reprinted[5] and gave me a chance to read the original texts, I came to the conclusion that my "physical explanation" of 1947 in essence was a *rediscovery* of the physical definition, which Ambartsumian had in mind to begin with, but which had been lost in the more mathematical formulation of Chandra. Chandra acknowledged this interpretation in his book.[6] However, I feel that he was never impressed by its full force, and presented this as an illustration, rather than as a basic definition, as I would have preferred. Later, in my own book[7] I gave what I feel is a simple but complete story.

Unfortunately, prior commitments prevented me from being present at the important meeting in Armenia[8] honoring Ambartsumian. At that meeting, Chandra gave an excellent review of his own work, obviously again with emphasis on the mathematical aspects.

What can I conclude from these episodes? Recounting them, I am struck by their similarity: two rather simple optical problems in ordinary three-dimensional space. Solving such a problem feels a little like being a rabbit nibbling a cabbage that happens to grow in the field. Tasty, but not a big achievement.

In contrast to that, Chandra spent 80 per cent of his active life working on problems in relativity and in quantum mechanics, and in other topics requiring his superior mathematical skill. They provided the food more fitting to a giraffe. After all, I feel that it would be silly to blame a giraffe for perhaps once overlooking a cabbage.

Notes

1. K.C. Wali, *Chandra: A Biography* (University of Chicago Press, 1991).
2. S. Chandrasekhar, *Truth and Beauty, Aesthetics and Motivation in Science* (University of Chicago Press, 1987).
3. H.C. van de Hulst, "Optics of Spherical Particles" (thesis, Utrecht, 1946).
4. H.C. van de Hulst, "Scattering in a Planetary Atmosphere," *Astrophys. J.* **107**, 220 (1948).
5. V.A. Ambartsumian, *Nauchni Trudi* (Scientific Works), V.V. Sobolev, editor, 2 volumes (Izd. Akad. Nauk Armyanskoy SSR, Yerevan, 1960).
6. S. Chandrasekhar, *Radiative Transfer* (Oxford Clarendon Press, 1950).
7. H.C. van de Hulst, *Multiple Light Scattering, Tables, Formulas and Applications*, 2 volumes (New York Acad. Press, 1980).
8. M.A. Mnatsakian and H.V. Pikichain (editors), *Principle of Invariance and Its Application* (proceedings of a symposium held at Buyarakan, 1981). Academy of Sciences of Armenian SSR, Yerevan, 1989.

List of Contributors

George Anastaplo is a law professor at Loyola University, Chicago. However, he belongs to more than one academic discipline, including liberal arts, political philosophy and political science. His lifelong dedication to the study of the American Constitution has resulted in several major books. The partial list of these books contains: *The Constitution, Human Being and Citizen, The Constitution of 1987, The Amendments to the Constitution* and *The American Moralist.*

Abhay Ashtekar is the Director of the Center for Gravitational Physics and Geometry, and Eberly Professor of Physics at Penn State University, University Park, Pennsylvania. He is well known for his numerous fundamental contributions to general relativity and quantum gravity, particularly the description of Einstein's theory in terms of a set of new variables which has come to be known as the Ashtekar variables.

Richard Askey is John Bascom Professor of Mathematics at the University of Wisconsin, a Fellow of the American Academy of Arts and Sciences and an honorary fellow of the Indian Academy of Sciences. He is an authority in the area of special functions with numerous research papers, the author of the book *Orthogonal Polynomials and Special Functions*, and the Editor of the *Collected Papers of Gabor Szego.*

Bimla Buti is a plasma physicist at the National Physical Laboratory, New Delhi, India. Her wide-ranging research interests include linear and non-linear stability of relativistic and non-relativistic magnetoplasmas, non-solitary waves, turbulent and chaotic processes in interplanetary plasmas, solar corona, terrestrial iono- and magneto-sphere, cometary environment and pulsars. She is a fellow of the American Physical Society, Indian National Science Academy and Third World Academy of Sciences. She has been a Visiting Scientist at UCLA and NASA centers.

Parasu Balakrishnan, younger brother of Chandra, is retired from the medical profession. While practicing medicine, his real love throughout his life has been literature. He has written voluminously in Tamil and English about the poet Kalidasa, and about Indian epics and cosmology.

Lalitha Chandrasekhar, about whom Chandra said:

> The full measure [of my indebtedness] cannot really be recorded: It is too deep and too all-persuasive. Let me then record simply that Lalitha has been the motivating source and strength of my life. Her support has been constant, unwavering, and sustained. And it has been my mainstay during times of stress and discouragement. She has shared my life; selfless, devoted, and ever-patient and waiting.
>
> — *Chandra: A Biography of S. Chandrasekhar*
> (University of Chicago Press, 1991), p. 11

James W. Cronin, University Professor Emeritus in the Department of Physics and the Enrico Fermi Institute at the University of Chicago. He shared a Nobel Prize with V. Fitch for the discovery of CP-violation. Currently he is the spokesperson for the International Pierre Auger Project to build a detector system capable of detecting high energy gamma rays from space.

R.H. Dalitz, after completing a term of Royal Society Professorship, is currently Emeritus Professor of Theoretical Physics at Oxford University, England. Prior to Oxford, Dalitz was a Professor in the Department of Physics and the Enrico Fermi Institute at the University of Chicago. A leading figure for over four decades in high energy physics research, he is well known for many fundamental contributions to elementary particle physics, the most significant one being the tau–theta puzzle in the analysis of K-meson decays in the fifties which led to the discovery of parity violation in weak interactions.

Donna Elbert resides in Williams Bay, Wisconsin, after her retirement as Administrative Assistant to the Department of Astronomy and Astrophysics at the University of Chicago.

Valeria Ferrari is an Associate Professor of Physics at the University of Rome, Italy. She is member of the Committee of the International Society on General Relativity and Gravitation.

Peter G.O. Freund is a Professor in the Department of Physics and the Enrico Fermi Institute at the University of Chicago. He received his doctorate from the University of Vienna (Austria) and is well known for his many contributions to theoretical and mathematical physics, which include hadronic dual models that gave birth to string theory, string theory itself, supersymmetry, magnetic monopoles and higher-dimensional theories. A fellow of the American Physical Society, he was recently awarded an honorary doctorate by the West University of Timisoara, Romania.

John L. Friedman is a professor in the physics department at the University of Wisconsin-Milwaukee. His wide-ranging research interests include structure and stability of rotating relativistic stars and topological aspects of classical and quantum gravity.

Radha Sarma Hegde, Chandra's niece, is a faculty member in the Department of Communications at Rutgers University, New Jersey, USA.

Agnes M. Herzberg is in the Department of Mathematics at Queen's University, Kingston, Canada. She is the editor of *Short Book Reviews*, a publication of the International Statistical Institute.

Van de Hulst is a well-known Dutch astronomer living in Leyden, the Netherlands. His prediction and the subsequent discovery of the 21-centimeter atomic hydrogen line played a significant role in radio astronomy and the study of the structures of galaxies. He is a recipient of the Eddington Medal (Royal Astronomical Society), the Henry Draper Medal (National Academy of Sciences) and the Rumford Medal (American Academy of Arts and Sciences).

Sanjay Kumar is a Delhi-based journalist and a documentary film maker. He is a frequent contributor to *New Scientist* and *The Lancet*. He is in the process of making a documentary on Chandrasekhar.

Norman Lebovitz is a Professor of Mathematics at the University of Chicago, from which he got his doctorate in physics under Chandra's direction. Lebovitz's research interests have centered on astrophysical fluid dynamics and on asymptotic solutions of ordinary differential equations. He has held visiting professorships at California Institute of Technology, Tel-Aviv, Punjab and Sussex Universities. He has also held fellowships from the Belgian-American Foundation, Caltech, the Sloan and the Guggenheim Foundations.

Anne Magnon is a professor in the department of mathematics in the Institut Universitaire de France at the Universitaire Blaise Pascal, Clermont-Fd, France. Noted for her researches in the general theory of relativity and cosmology, she has translated *Chandra: A Biography of S. Chandrasekhar* into French (published by Frontieres).

Leon Mestel, a Fellow of the Royal Society, London, is Emeritus Professor of Astronomy at the University of Sussex, England. A former Fellow of St. John's College, Cambridge, and Professor of Applied Mathematics, Manchester, he is co-author of the book *Magnetohydrodynamics* with N.O. Weiss.

Jayant V. Narlikar is the Director of the Inter-University Centre for Astronomy and Astrophysics (IUCAA), Pune, India. He is a leading proponent of quasi-steady-state cosmology as an alternative to the popularly accepted big bang cosmology. He is the author of several books, including *The Lighter Side of Gravity* and *The Structure of the Universe.*

Takeshi Oka is a professor in the Departments of Chemistry and Astronomy and Astrophysics at the University of Chicago.

Eugene Parker is S. Chandrasekhar Distinguished Service Professor in the Departments of Physics and Astronomy and Astrophysics, and in the Enrico Fermi Institute at the University of Chicago.

Roger Penrose is Rouse Ball Professor of Mathematics at the University of Oxford. Well known for many of his fundamental contributions to the general theory of relativity, he won the Wolf Prize in 1988, which he shared with Stephen Hawking. Penrose is the widely acclaimed author of *The Emperor's New Mind* and its sequel, *Shadows of the Mind*, which offer a grand survey of modern physics and in its light explore the nature of the human mind and human consciousness.

S. Ramaseshan, a distinguished scientist, is a founder of many scientific institutions in India. He was the President of the Indian Academy of Sciences and editor of the scientific journals *Pramana* and *Current Science.* He has edited the collected scientific papers of C.V. Raman.

Martin Rees is Royal Society Research Professor and Astronomer Royal. He was previously Plumian Professor of Astronomy and Experimental Philosophy and Director of the Institute of Astronomy at Cambridge University. He is a leading figure in astrophysics, and a member of the Royal Society and several foreign academies. His books include *Perspectives in Astrophysical Cosmology* (Cambridge University Press, 1995), *Gravity's Fatal Attraction: The Discovery of Black Holes* (W.H. Freeman, 1995; with M.C. Begelman) and *Before the*

Beginning: Our Universe and Others (Simon and Schuster, 1997).

Robert G. Sachs is Emeritus Professor in the Department of Physics and the Enrico Fermi Institute at the University of Chicago. Aside from his many fundamental contributions to nuclear and elementary particle physics, he has played a leading role in science and energy policies of the United States. He was the Director of the Argonne National Laboratory and the Enrico Fermi Institute.

Rafael Dolnick Sorkin grew up in Chicago and was educated at New Trier Township High School, Harvard University and the California Institute of Technology. Currently he is professor of physics at Syracuse University on leave at ICN-UNAM in Mexico City. His research is directed broadly to the problem of reconciling general relativity with quantum mechanics, and his current efforts in that direction center on a theory which replaces continuous spacetime by a collection of discrete "atoms" structured as a locally finite order. Beyond science, he takes interest in music, languages and political economy.

Stephen M. Stigler is Ernest Dewitt Burton Distinguished Service Professor in the Department of Statistics at the University of Chicago. He has written extensively on the history of statistics, including a 1986 book that discussed the role of problems in astronomy and geodesy in the early development of statistical methods, *The History of Statistics: The Measurement of Uncertainty Before 1900.* Stigler holds an appointment in the Committee for the Conceptual Foundations of Science.

Noel M. Swerdlow is a Professor in the Departments of Astronomy and Astrophysics and History, and a member of the Committee for the Conceptual Foundations of Science. He is an eminent Copernican scholar and a historian of ancient astronomy.

V.L. Telegdi, a charismatic and leading figure in high energy physics, received the internationally prestigious Wolf Prize in 1991 for his outstanding researches. He lives in Geneva, and has been associated with CERN after retiring from ETH, Zurich. Prior to that he was a Professor in the Department of Physics and the Enrico Fermi Institute at the University of Chicago.

Saul A. Teukolsky, a leading astrophysicist, is a Professor of Physics and Astronomy at Cornell University, Ithaca, New York, USA. His research interests include general relativity, relativistic astrophysics and computational physics. His earliest work led to the development of the "Teukolsky equation," which describes how a black hole interacts with surrounding objects. His subsequent research has included the physics of pulsars and supernova explosions, properties of rapidly rotating stars, stellar dynamics, and planets around pulsars. Currently he is among a group of scientists investigating the use of high performance computing to solve Einstein's equations of general relativity. He is co-author of several widely used textbooks.

Roza Michalska Trautman and *Andrzej Trautman* received their doctorates in physics while working in the research group of Leopold Infeld in Warsaw, Poland. Andrzej Trautman is a Professor and Head of the Chair for Relativity and Gravitation at the Institute of Theoretical Physics of Warsaw University. Well known for many fundamental contributions to general relativity, he is the author of four books on general relativity, applications of differential geometry to physics and spinors. Roza Trautman has been co-author with Infeld of several papers on gravitational radiation; she has also worked on non-linear problems in quantum optics at the Institute of Physics of the Polish Academy of Sciences.

Vatsala Vedantam was an assistant editor of the *Deccan Herald*, Bangalore. Currently she is a free-lance journalist writing for select magazines and newspapers in India and abroad. The account of her interview in this book is condensed from her article in the *Deccan Herald*, with the permission of its editor.

Robert M. Wald is a Professor in the Department of Physics and the Enrico Fermi Institute at the University of Chicago. A leading researcher in relativity, he is the author of the widely acclaimed book *General Relativity*, the first textbook on the subject with a totally modern point of view. He is also the author of *Space, Time and Gravity, The Theory of the Big Bang and Black Holes* and, more recently, *Quantum Field Theory in Curved Spacetime and Black Hole Thermodynamics.*

Kameshwar C. Wali is a Professor of Physics at Syracuse University, specializing in elementary particle physics. He is a Fellow of the American Physical Society and a founding member of the Forum of History of Physics. He is the author of *Chandra: A Biography of S. Chandrasekhar* (University of Chicago Press, Chicago, 1991).

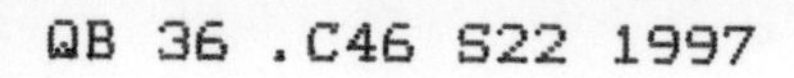